Wood Heat

A Practical Guide to Heating Your Home with Wood

Andrew Jones

FIREFLY BOOKS

Published by Firefly Books Ltd. 2014

First printing

Publisher Cataloging-in-Publication Data (U.S.)
Jones, Andrew.
Wood heat : a practical guide to heating your home with wood / Andrew Jones.
[192] pages : col. ill. ; cm.
Includes references and index.
Summary: This resource for wood heating in the home covers: the selection of types of wood, the science of burning, types of heaters, the chimney, the optimal location for the heat appliances and more.
ISBN-13: 978-177085-299-0 (pbk.)
1. Fuelwood. 2. Heating – Equipment and supplies. 3. Dwellings – Heating and ventilation. I. Title.
662.65 dc 23 TP324.J654 2014

Library and Archives Canada Cataloguing in Publication
Jones, Andrew, 1963-, author
Wood heat : a practical guide to heating your home with wood / Andrew Jones.
Includes references and index.
ISBN 978-1-77085-299-0 (pbk.)
1. Fuelwood. 2. Heating--Equipment and supplies.
3. Dwellings--Heating and ventilation. I. Title.
TP324.J66 2014 662.6'5 C2014-901160-1

Published in the United States by
Firefly Books (U.S.) Inc.
P.O. Box 1338, Ellicott Station
Buffalo, New York 14205

Published in Canada by
Firefly Books Ltd.
50 Staples Avenue, Unit 1
Richmond Hill, Ontario L4B 0A7

Cover and interior design: Hartley Millson
Illustrations: Clive Dobson
Maps: George A. Walker

Printed in China

The publisher gratefully acknowledges the financial support for our publishing program by the Government of Canada through the Canada Book Fund as administered by the Department of Canadian Heritage.

Acknowledgments

I would like to thank my editors, Michael Mouland and Tracy C. Read, for their support and deftness with my words. This book is dedicated to my parents, keepers of the flame.

Contents

Introduction

Why Wood?

IN THIS DAY AND AGE, IS IT REASONABLE TO HEAT with wood when more efficient sources of energy exist? Is wood really cheaper? Is the wood-burning lifestyle for you? This book seeks to answer these questions. It explores everything from how to source, choose and purchase wood and how to season, split, stack and store it to how to build and tend fires. It explains the differences between wood-burning appliances, and it provides advice on how to heat your residence safely, whether you are warming your living room or your entire home.

Those who've had the experience would agree that there are few things more comforting in life than basking in the warmth of a crackling wood fire as the day draws to a close. Yet it's also true that heating with wood provides benefits which extend well beyond basic comfort and warmth: Wood is a sustainable fuel that can be harvested ethically and burned cleanly, and doing so can result in a great sense of personal satisfaction and well-being.

But if wood is a relatively inexpensive, renewable source of energy with such tremendous potential, why don't more people heat with it? Why isn't it part of energy policy discussions by local, regional and national governments? Why is heating with wood still largely seen as a quaint backwoods pursuit?

The answer to the first question is practical. Home heating with wood on an urban scale is unrealistic, for reasons that start with basic distribution and space. Unlike other renewable sources of energy that come straight to your house, such as electricity and natural gas, firewood requires a lot of dedicated space if you are to use it to warm a single-family dwelling through a frigid northeastern winter.

Another drawback to heating with wood is air pollution. Wood that is not burned properly creates noxious smoke, and as a result, it is strictly regulated in many municipalities throughout Canada and the United States. Older stoves and fireplaces that don't burn wood as cleanly and efficiently as newer appliances can exacerbate this problem, creating more smoke. Burning wood is thus often seen as a nuisance rather than a beneficial resource.

Yet when all things are considered, the advantages of heating with wood in some settings can handily outweigh the disadvantages, as we'll learn. Wood is a sustainable resource that requires little processing, one that limits your furnace use and offers self-sufficiency when the power goes out. Sourcing and buying wood locally also keeps your energy dollars within the community and contributes to responsible resource management.

Make no mistake—there is a considerable financial, physical and lifestyle investment in deciding to heat a home with wood. This book is not for everyone. While many will appreciate the warm feeling heating with wood evokes, fewer will take the steps required to make it a reality. But those who do will be rewarded handsomely.

A Short History of Wood Burning

Humans have burned wood to keep warm for millennia. What many of us don't realize, however, is that the use of wood as our primary home-heating fuel was displaced only relatively recently. Indeed, residential wood burning was the main source of domestic heat throughout Canada and the United States until about 150 years ago, and for much of rural North America, wood was virtually the only practical heating-fuel option until after the Second World War.

After 1945, the use of firewood for home heating dropped rapidly. Many associated wood heating with rural poverty and hardship and embraced the widespread, convenient and inexpensive energy sources of the future. During the 1950s and 1960s, the practice of wood burning was largely relegated to decorative fireplaces installed in new single-family homes that were heated with oil, natural gas and electricity. During these pivotal decades, wood became a fuel of the past. Eventually, it came to be viewed mostly as a fuel for recreational use.

It took the energy crisis of the 1970s to rekindle interest in wood heating in North America. The triple threat of skyrocketing oil prices, high interest rates and the recession of the early 1980s put the squeeze on household budgets. For many families, switching to wood became the simplest way to stop the financial bleeding caused by soaring monthly home-heating bills. Suddenly, the fusty old woodstove became a symbol of Canadian and American resourcefulness, ingenuity and self-sufficiency.

The resurgence of the woodstove, however, created an entirely new set of problems. Until U.S. Environmental Protection Agency (EPA) regulations came into effect in the late 1980s (followed by Canadian Standards Association (CSA) standards in 2000), wood-burning technology had been fairly crude. Most wood-burning appliances were little more than steel or cast-iron boxes in which to build a fire. Old cast-iron stoves and furnaces burned inefficiently, sending acrid smoke through long, leaky flue pipes to crumbling, unlined masonry chimneys. Gaskets on loading doors were rare. An entire continent rushed back to wood heating after decades of inattention and decline, and a generation of people with no collective memory of the dangers of house fires led to thousands of new wood-burning converts losing their homes. Wood heating suddenly became one of the leading causes of residential structural fires, and many people to this day retain the notion that wood heating is inherently hazardous.

Throughout the 1980s, a concerted effort by all levels of government, the insurance industry and the wood-heating industry resulted in the development of safety standards and professional training programs. By the middle of the decade, the price of oil and other energy commodities had settled back to manageable levels, and the economy had recovered. Gradually, wood heating once again lost its glow, and the new and improved gas fireplace swiftly supplanted the

10 Benefits to Burning Wood

1. **Wood Renews Itself Faster.** Fossil fuels such as oil and natural gas are technically renewable energy resources, but trees can be planted to replace themselves much faster than new oil and gas can be made. While wood could never supply the world's heating needs, responsible planting programs and forestry stewardship not only could lead to an inexhaustible supply of firewood but could ease the strain on traditional energy sources.

2. **Cut the Energy Cord.** Tired of writing monthly checks to monolithic energy companies that have no compunction about limiting service or raising rates? Heating with wood allows you to cut the energy cord with these corporations and puts you in the driver's seat. You get self-sufficiency as well as a warm hearth.

3. **Bye-Bye Blackouts.** Those big utilities are not infallible. In fact, they're often far from reliable. When a major storm cuts off the power, as it did across southern Canada and the northern United States in December 2013, most conventional heating systems are dead in the water: Electrical baseboards cool, expensive heat pumps fall idle, gas furnaces are sidelined. But a woodstove or fireplace can keep you warm, cozy and safe and allows you to offer refuge to any of your neighbors still reliant on the grid for their heat.

4. **Heat Selectively, Save Across the Board.** Well-planned space heating with wood can save you even more on energy costs. Put a wood-burning stove or fireplace in the family room, where you spend the most time, and keep the basement and bedrooms cool. Regardless of your current choice of fuel, this approach can help you save a quarter of your annual home-heating costs.

5. **Keep It Local, Pay It Forward.** Burning with wood keeps your energy dollars invested in local businesses. And if a majority of residents heat with wood, the resultant pool of savings effectively means more disposable income that stays and circulates within the community.

6. **Burning Wood Warms Your House, Not the Atmosphere.** Young trees *absorb* carbon dioxide as they grow—so unlike oil and natural gas pumped from the depths of the planet (whose emission of carbon dioxide *adds* to the total load of carbon dioxide released into the atmosphere), burning wood is greenhouse gas neutral, as what is released is eventually recycled by growing trees.

7. **Responsible Resource Management.** Many woodlots practice an admirable model for sustainable forestry. By heating your home with wood, you're supporting woodlot owners who responsibly thin forests, clear out fallen and dead trees and create room for new saplings to grow, thereby providing a source of heat for generations to come.

8. **A Little Ray of Sunshine.** Trees store energy drawn from the sun as they grow. Burning wood unleashes this stored energy in the form of crackling flames and radiant heat. In the dead of January, this is like a sunny day saved and unfurled in your living room, warming you with the heat of summers past.

9. **Fire, Fire Burning Bright.** Charlie Brown once famously said that there are three things in life people like to stare at: a flowing stream, a crackling fire and a Zamboni clearing the ice. Indeed, a roaring fire can offer unmatched solace and comfort. A fire's pool of warmth and amber glow invite intimate conversation, family gatherings or solitary reflection.

10. **You'll Really Feel the Burn.** Last, but certainly not least: If you use wood as the primary source of heat in your home, you'll never be in better physical shape. Hauling, chopping, splitting and stacking wood are not for the faint of heart and will form part of a rigorous fitness regime, particularly during our long northern winters!

woodstove. Even with the stock market crash of 1987, convenience trumped energy self-reliance for many.

Then came 1998. In January, a crippling ice storm shut down much of eastern Canada and the northeastern United States for weeks, and once again, the lowly woodstove and the wood-burning furnace saved the day for those lucky enough to own one. The following autumn, woodstove dealers could not keep up with demand by new customers, who didn't want to risk another winter without a solid home-heating backup plan.

That same year, geologists Colin Campbell and Jean Laherrère published an article in *Scientific American* titled "The End of Cheap

Oil." In it, they broached the notion of "peak oil" for the first time. The pair accurately foretold that the global peak of oil production was near. Once that peak was reached, the price of oil would begin to rise steadily and never fall again. Wood burning began to enjoy its second renaissance.

The good news is that by the turn of the 20th century, heating with wood had evolved dramatically. It is now safer, more efficient and less taxing. Where burning with wood once meant backbreaking work, crude stoves and dangerous chimneys, now portable chainsaws cut the work in half, the efficiency of the average woodstove has roughly doubled, to around 70 percent, chimney technology and safety have vastly improved, and rigorous building codes and recognized standards have been established for virtually every type of wood-burning appliance.

Wood-heating retailers, contractors, chimney sweeps and municipal and insurance inspectors now benefit from national training and certification programs. Trained professionals can rely on clear and comprehensive safety rules. Public information, much of it produced by both governments and industry groups, is putting to rest old fears about the dangers of wood burning.

As recently as the mid-1980s, wood burning was relegated to a basement wood furnace or a stout cast-iron woodstove. Today, heating with wood has become retro chic as many North Americans try to recapture memories of their childhood as they refit cottages or build rural getaways. Woodstoves, by far the most popular type of wood-burning appliance, are no longer stodgy black boxes but are available in appealing enamel colors and both traditional and modern designs. Pellet stoves can provide 24 hours of heating in a single charge. Safe fireplace inserts allow wood to be burned cleaner at higher efficiencies. Breakthroughs in glass composition have resulted in the manufacture of doors that allow efficient heating without hiding the beauty of the flames.

By fusing innovation with tradition, wood burning is at last in a position to offer the best of the past and the future.

Chapter 1

Who Burns Wood?

IN A WORLD OF TOUCH-SCREEN SMARTPHONES, web-enabled refrigerators and retina scanners, wood heating might seem resolutely rough, hands-on, back-to-basics stuff. Are people who heat with wood out of step with the modern world, as it hums around them through fiber-optic cables and along eight-lane expressways? Has the wood-burning community simply withdrawn and settled comfortably off the grid? Can we learn something from the quiet path they tread?

We do know that more than a decade into the 21st century, wood remains an important residential energy resource in North America, especially outside large urban areas. The numbers might surprise you. More than 10 million U.S. households use wood as their main heating fuel or to supplement other heating fuels. Over 25 percent of Canadian households—that's one in four—still burn wood.

Take a drive through the rural back roads of your province or state, down country lanes bordered by thickets of forest, and you won't get far without seeing long lines of pale, piled firewood standing silent in yards. This wood, cut from woodlots in winter and split and stacked in spring to dry in the summer sun, is eventually moved into the house, where it will keep families warm in winter in a seasonal cycle as old as the early settlers who first built fires in their homes.

The largely rural firewood fraternity has seemingly figured out a way to reduce greenhouse gases and the pressure on dwindling fossil fuels, keep home-heating costs down and strengthen their local communities. They have made firewood the ultimate populist energy resource.

However, there *are* dissenting voices—activists clamoring to have wood burning banned from their communities because of air pollution. Some environmentalists, fearing negative impacts on our forests, have also sounded alarm bells over the increased use of wood burning.

Environmental Impact

Wood is not a perfect fuel, but does such a thing exist? All energy generation and consumption by nature creates unwanted by-products, and impassioned arguments about the legacies of coal and uranium mining and natural gas and oil extraction are ongoing. Even famously "clean" energy sources like wind and solar power have been found to have a detrimental impact on the environment.

The biggest drawback and major environmental impact of wood burning is, of course, visible for all to see—wood-smoke pollution. Three aspects of this pollution are discussed and debated, sometimes hotly: nuisance smoke (caused by neighbors inefficiently heating their homes); air-shed contamination (caused by too much smoke produced in areas with a depressed topography, such as a river valley, which is prone to temperature inversions in the winter that trap smoke close to the ground); and indoor air pollution (caused by leaky or inefficient in-house wood-burning appliances).

There is no getting away from the fact that burning wood produces smoke, which is unhealthy to breathe in high concentrations. Even in low concentrations, it can be harmful to children, the elderly and people with lung diseases or allergies. Those who would like to curb wood burning in some communities point out that wood smoke contains toxic chemicals like polycyclic aromatic hydrocarbons and dioxins as well as known carcinogens such as furans and acrolein. This despite the fact that a poorly fired backyard barbecue or a city bus can emit equally noxious substances.

Table 1.1

Energy Content in Megajoules (MJ) and Local Price of Various Fuels		
Energy Source	**Energy Content**	**Your Local Price**
Oil	38.2 MJ/L	/L
Electricity	3.6 MJ/kWh	/kWh
Natural gas	37.5 MJ/m³	/m³
Propane	25.3 MJ/L	/L
Hardwood (air-dried)	30,600 MJ/cord	/cord
Softwood (air-dried)	18,700 MJ/cord	/cord
Mixed fuelwood (air-dried)	25,000 MJ/cord	/cord
Wood pellets	19,800 MJ/tonne	/tonne

Source: *A Guide to Residential Wood Heating*, CMHC (2008), p.79

The good news is that the new wood-burning technology built into today's certified stoves (see "Certified Wood-Burning Appliances" on p. 56) goes a long way toward addressing all three aspects of wood-smoke pollution. Thanks to the four decades of research since the 1970s oil crisis, we know a great deal more about wood burning. Advanced-technology stoves, inserts, fireplaces and furnaces can reduce wood smoke by up to 90 percent when compared with older, so-called airtight stoves.

For example, older woodstoves emit, on average, at least 25 grams of smoke per hour of operation (g/h), while the emissions from older wood-fired outdoor boilers can range between 50 g/h to 100 g/h. In contrast, both U.S. Environmental Protection Agency (EPA) regulations and the Canadian Standards Association's *Performance Testing of Solid-Fuel-Burning Heating Appliances* (CSA B415.1) limit emissions of certified non-catalytic wood-burning stoves to no more than 4.5 g/h.

Table 1.2

Typical Seasonal Heating System Efficiencies		
Fuel	**Type of System**	**Efficiency**
Oil (furnace or boiler)	Cast-iron head burner	60%
	Retention head burner	70%–78%
	Mid-efficiency furnace or boiler	83%–89%
Electricity	Furnace/boiler or baseboard	100%
	Geothermal (ground-source heat pump)	260%
Natural Gas	Furnace/boiler (conventional)	55%–65%
	Furnace/boiler (mid-efficiency)	78%–84%
	Furnace/boiler (condensing)	90%–97%
Propane	Furnace/boiler (conventional)	55%–65%
	Furnace/boiler (mid-efficiency)	79%–85%
	Furnace/boiler (condensing)	88%–95%
Wood (conventional)	Furnace/boiler	45%–55%
	Advanced* furnace/boiler	55%–65%
	Conventional stove†	50%–65%
	Advanced* stove†	65%–80%
	Advanced* fireplace	50%–70%
	Pellet stove	55%–80%

*Refers to appliances certified as low-emission by the EPA or adhering to the CSA B415.1 standard

†Properly located

Source: *A Guide to Residential Wood Heating*, CMHC (2008), p.79

Since these regulations were first established in 2000, the average emissions of certified stoves have declined steadily due to advances in technology and healthy competition between manufacturers. Many woodstove models on the market today emit only 2 g/h to 4 g/h.

By burning and not wasting the energy-rich smoke, these stoves increase the efficiency of providing heat to the home. Older stoves were grossly inefficient, with efficiency rates as low as 35 percent for a cast-iron box stove to 55 percent for a so-called airtight model in the 1970s. CSA- and EPA-certified woodstoves spill less smoke into the indoor air because fires don't tend to smolder in them, with the average being around 70 percent efficient. Even the new breed of emissions-certified outdoor boilers are much more efficient (see "Outdoor Boiler" on p. 74).

This efficiency translates directly into less smoke going up the flue and minimizes the telltale gray clouds billowing from chimneys. Multiply that across a small community of wood-burning homes, and both nuisance smoke and air-shed contamination are taken care of in one stroke.

So wood burning's environmental triple threat—air-shed contamination, nuisance neighbors and indoor air pollution—can be addressed by upgrading home and wood-burning equipment to the latest efficiency standards, launching public information and awareness campaigns and, where necessary, updating regulations.

Sustainability

Is wood heating environmentally sustainable? The answer to this question hinges on two contentious issues: greenhouse gas emissions and deforestation. The oft-cited argument that burning wood increases the amount of carbon dioxide going into the atmosphere tells only half the story. The carbon dioxide would, in any case, be emitted by these trees when they die, and young trees actually *absorb* carbon dioxide, rendering wood-burning greenhouse gas neutral and the net carbon dioxide emissions far below those of fossil fuels such as oil and natural gas, which are extracted from deep beneath the Earth's surface and whose carbon dioxide emissions add to greenhouse gas in the atmosphere.

As for deforestation, large parts of Canada and the northern United States are relatively thinly populated and have highly productive forests. These are the regions where wood heating makes sense. Foresters have claimed wood burning could double in most of these regions without unduly straining forestry resources, as long as selective harvesting is used to thin dense stands and remove poor-quality trees, and seed trees of a wide variety of species and ages are left standing.

Indeed, the careful work by generations of farmers and other woodlot owners, visible in healthy, productive woodlots that have consistently produced wood for home heating, provides a blueprint for stewardship that the entire forest industry can follow.

It has often been said that a healthy, well-managed woodlot can yield a cord of wood per 2½ acres (1 ha) per year forever (see "Just How Much Is a Cord, Anyway?" on p. 34) and that a 12-acre (5 ha) woodlot could sustainably produce enough firewood each year to fully heat a house. While true as a rule of thumb, it needs a little updating. It takes a lot less than 5 cords of wood to heat a new energy-efficient house using a modern wood-burning stove. Indeed, with the efficiencies offered by today's wood-burning appliances, additional homes could be heated from the same woodlot.

Yet despite its considerable advantages, firewood is not the magic bullet that will solve the problems of global warming. Heating with wood is not a meaningful option for all homes, such as multi-family dwellings in urban downtown cores, where regulations forbid it and where the air is already choked with pollution from industry and transportation. Finally, as we mentioned earlier, a winter's worth of wood takes up a lot of space, and feeding fires all winter requires a heightened level of physical fitness and attentiveness. Clearly, wood heating is not for everyone.

But for those who live outside major cities, in small towns and rural areas, where both population density and the cost of firewood are lower, and for those who are looking to offset their monthly home-heating bill and provide some nostalgic warmth in the bargain, it just might be worth honing up on your chopping, stacking and kindling skills.

Table 1.3

Typical Heating Loads in Gigajoules (GJ) for Canadian Detached Homes	
City	**Heating Load**
Victoria/ Vancouver, B.C.	60
Calgary, Alta.	90
Edmonton, Alta.	95
Regina/ Saskatoon, Sask.	90
Winnipeg, Man.	90
Yellowknife, N.W.T.	145
Thunder Bay, Ont.	95
Ottawa, Ont.	75
Toronto, Ont.	65
Montréal, Que.	80
Quebec City, Que.	85
Saint John, N.B.	75
Charlottetown, P.E.I.	80
Halifax, N.S.	75
St. John's, N.L.	85

Source: *A Guide to Residential Wood Heating*, CMHC (2008), p.79

Table 1.4

Typical Heating Loads in Gigajoules (GJ) for U.S. Detached Homes	
City	**Heating Load**
Albuquerque	12.8
Atlanta	14.3
Boston	46.5
Chicago	93.0
Denver	75.2
Fort Worth	10.1
Kansas	85.2
Los Angeles	1.5
Miami	0.2
Minneapolis	156.1
New Orleans	4.9
New York	70.2
Phoenix	9.4
San Francisco	7.8
Seattle	24.6
Washington	48.2

Source: Franco, V., Lutz, J., Lekov, A., Gu, L., "Furnace Blower Electricity: National and Regional Savings Potential," Publication FSEC-CR-1774-08 (Florida Solar Energy Center 2008)

Switching to Wood: Comparing Annual Heating Costs

Is it financially worth your while to heat with wood? Whether you're thinking of upgrading to a new, higher-efficiency wood-burning appliance or converting entirely to wood, you can use the following charts and calculations to determine the relative cost of heating with wood in your area compared with the fuel you are currently using to heat your house.

1. First, call your local fuel supplier to find out the average cost of heating with your current energy source, and fill in that cost in Table 1.1, using the units of measurement provided. Then call local wood suppliers to do the same, keeping in mind that natural firewood is measured in full 4-by-4-by-8-foot (1.2x1.2x2.4 m) cords (see "Just How Much Is a Cord, Anyway?" on p. 34).
2. Next, select the city closest to you from Table 1.3 or 1.4.
3. Use the formula below to calculate the annual heating cost:

Sample calculation: Grant lives near Saskatoon, Saskatchewan, in a new detached home with a heating load of 70 gigajoules that he heats with an oil furnace which is 80 percent efficient. Oil costs, on average, around $1.28 per liter. He is considering switching to a 75 percent efficient advanced-technology woodstove and heating with air-dried hardwood, which is sold in his area for $320 per cord.

Existing heating system:
(1.28 / 38.2) x (70 / 80) x 100,000 = $2,931.94

Proposed heating system:
(320 / 30,600) x (70 / 75) x 100,000 = $976.03

By switching entirely to wood, Grant could save close to $2,000 on his annual heating costs. Of course, these are comparative guidelines, and many other variables can come into play, such as the percentage of your house you heat. A good guide is to determine which neighbors in your area heat with wood. Look for telltale woodpiles in yards, and strike up a friendly conversation. Direct advice is always better than a formula.

$$\frac{\text{Energy Cost/Unit (Table 1.1)}}{\text{Energy Content (Table 1.1)}} \times \frac{\text{Heating Load (Table 1.3 or 1.4)}}{\text{System Efficiency (Table 1.2)}} \times 100{,}000 = \text{Approximate Annual Heating Cost}$$

The Relative Efficiency of Wood Burning

Firewood is typically consumed in three ways: It is burned in a decorative fireplace; it is used as a supplementary energy source to heat a basement, a family room or an extension to the house; or it serves as a primary heating source. The first is likely to occur in cities, the second is the largest segment of wood use for home heating, and the third is commonplace in rural areas.

If you are considering the second or third scenario, you may well ask whether wood heating is really economically viable in this day and age. The quick answer is yes. Dollar for dollar, wood burning is a cheaper way to heat your home than are conventional fuel sources such as oil, natural gas and electricity. Exactly how much cheaper depends on the type of home you want to heat, where you live and the local cost of various fuel types. To calculate how much you could potentially save by heating with wood, see "Switching to Wood: Comparing Annual Heating Costs" on p. 24.

These models serve as a rough guide only and address heating the entire house to an even temperature throughout. Maybe you want to heat just a single room with a woodstove. Even in that case, switching to wood burning could reduce the energy needed to heat this room by as much as 25 percent.

The real answer, however, is more complex and lies in how willing you are to adopt a wood-burning lifestyle. Monetary costs are only the beginning. Heating with wood is hands-on, intensive and messy. Consider the following questions before you decide. Do you have the space outside to cover and store a winter's supply of firewood? Do you have space inside your house to accommodate a few days' supply in anticipation of those winter storms that close down the roads? Do you have the physical strength and stamina required to split, move and stack firewood on a regular basis? Do you have time to devote to managing your newfangled fuel supply, tend fires and deal with new maintenance tasks like hauling out ashes? Can you put up with wood chips scattered throughout your house?

If your answer to these questions is, for the most part, yes, then you have what it takes to make the switch.

The Value and Satisfaction of Heating with Wood

Heating with wood is about a lot more than warming your home. It is a tangible expression of self-reliance, of the courage it takes to buck the trends and resist the lure of a sedentary, always-connected existence. Producing fuel to heat the family home during a long, cold northern winter is an accomplishment with significant meaning for those who are up to the task. The sense of personal commitment to social and environmental responsibility it can provide is powerful but rarely seen in the mainstream media, where bike marathons for charitable causes rule the day. Heating with wood will teach you more about the ethical cause-and-effect relationships of energy production and consumption than simply paying a utility bill. Heating with wood reinforces links to the land and represents a willing submission to the cycle of the seasons. It provides stability and security in a turbulent world.

Chapter 2

Wood

AN ARBORIST MIGHT TELL YOU THAT THE innermost part of a tree—the heartwood—has long stopped growing and that the only truly living layer in a tree's trunk is the cambium, the microscopically thin layer of green cells between the bark and sapwood. Anyone who has enjoyed a roaring fire, however, would heartily dispute that notion, as wood is never more alive than when it is being consumed in a blaze of hot energy. We treat fire as a living thing, saying that we have "breathed life into a fire" or that it has "died out." Admiration for the beauty of wood in all its forms is one of the cornerstones of the wood-burning lifestyle, and those drawn to it will find endless hours of fascination with this most beautiful of fuels. As William Blake once famously said, "I can look at the knot in a piece of wood until it frightens me."

Buying Wood

Buying firewood can be one of the most pleasurable activities associated with heating your home with wood. A trip to the country woodlot to inspect the neatly stacked logs, admiring patterns in the bark and wood, trading tips with fellow wood burners, selecting a mix of species for different burning needs—all contribute to the first step in your journey to becoming self-sufficient in the realm of home heat. It's possible to choose woods according to fire type, burning time, heating characteristics, aroma and aesthetics, and there are species for all occasions (see "Which Wood Is Best?" on p. 33).

Sourcing and purchasing *good* firewood can be tricky, however, as it is not regulated in all provinces and states throughout North America. If you're not careful, you could end up with a truckload of green wood that is impossible to burn cleanly in the upcoming heating season. The more time and effort you put into selecting a reliable supplier of firewood, the more this groundwork will pay off over the long term.

When looking for a firewood supplier, keep the following tips in mind:

- Ask neighbors who heat with wood to recommend someone reliable.
- Shop around and compare prices.
- Look for an established supplier who employs responsible forest-management practices—for instance, thinning rather than clear-cutting.
- Never order wood by telephone. Go in person to inspect the wood, and bring a tape measure to check piece length and pile size.
- Ask the supplier when the wood was felled and split. Has it been stacked off the ground and under cover? Are the logs hardwood, softwood or a mixture? Where is it from and is it a sustainable source?
- Look for wood that is clean, truly seasoned and neatly stacked. Do not buy randomly piled wood.
- Is the wood priced by weight or by volume? Selling by weight is problematic if you are buying it green to store until it's dry. You'll be buying more moisture than wood.
- Can the wood be split prior to delivery?
- How will it be delivered?

As for which is the best type of wood to burn, all wood, regardless of tree species, is similar chemically and has roughly the same energy

Table 2.1

Heat Values and Fuel Equivalents of Various Wood Species

	Wood (1 air-dried standard cord)	Available heat (million BTUs)	Fuel Equivalents		
			Fuel oil (L)	Natural gas (m^3)	Electricity (kWh)
Hardest	Elm, rock	32.0	553	493	3,800
	Hickory, shagbark	30.6			
	Oak, white	30.6			
	Locust, black	28.5			
	Maple, sugar	29.0	503	453	3,500
	Beech, American	27.8			
	Oak, red	27.3			
	Birch, yellow	26.2			
	Ash, white	25.0			
	Maple, red	24.0	432	385	3,000
	Tamarack	24.0			
	Cherry, black	23.5			
	Birch, white	23.4			
	Walnut, black	19.5	326	289	2,200
	Hemlock	17.9			
	Aspen	17.7			
	Spruce	16.2			
	Pine, eastern white	17.1	276	246	1,900
	Basswood	17.0			
	Cedar, white	16.3			
Softest	Fir, balsam	15.5			

Source: www.ontariowoodlot.com/pdf_older/by_the_cord.pdf, www.tdc.ca/wood.htm

content per pound or kilogram (see Table 2.1 "Heat Values and Fuel Equivalents of Various Wood Species" above). Where species differ is in their density, which is why hardwoods, such as oak, maple and beech, release more heat per fire and burn longer than softwoods, like pine, spruce and cedar. Indeed, hardwoods have been the traditional favorite wood for burning throughout central and eastern North America for precisely these reasons.

But in many regions of North America today, hardwoods are no longer plentiful enough to allow their use as fuel; as a result, softwoods are increasingly part of the mix. Often seen as second-class firewood because they are slow-drying, are full of resin and sap and tend to throw sparks, softwoods provide excellent fuel for spring and fall use, when heat demand is lower. Advanced-technology wood-burning appliances, fireplaces and furnaces tightly control the combustion process and can therefore handle a wider variety of wood species than their conventional predecessors. As a result, these appliances get along with softwoods just fine. If there is any doubt left in your mind, remember that softwoods do an excellent job of heating homes in some of the coldest parts of Canada and the United States, where only spruce, pine and light deciduous trees like birch and aspen dominate the forests. Ultimately, it is more important to have wood that is cut and split to the right size and properly seasoned than it is to get the hardest wood available.

If you are buying firewood for an open fireplace, steer clear of spark-throwers like tamarack, pine and chestnut and species that produce sharp, strong smoke, such as walnut. With softer, faster-burning woods like poplar and birch, it's also hard to get a good roaring fire going. If you have a wood-burning stove, furnace or boiler, sparks are not an issue and you can combine softwood and hardwood to get a pleasing balance of flames and embers. By purchasing a mix of species, you can contribute to sustainable woodlot management in the bargain, as some of these woods can be drawn from less desirable, dying and damaged trees.

Above all, your firewood must fit into your stove, furnace or fireplace. Pieces so large that they have to be jammed into the firebox make stoking the fire frustrating, if not impossible. For wood-burning stoves and furnaces, pieces should be at least 3 inches (7.5 cm) shorter than the size of the firebox. Even if your stove is big enough to take pieces as long as 20 inches (50 cm), shorter pieces are more desirable for ease of handling and fire maintenance. Good-quality firewood is cut into lengths that are roughly equivalent.

How much firewood is enough? Only experience can tell you how much you need to last the winter. Typically, a medium-sized modern

Which Wood Is Best?

FIRST CHOICE

- **Oak.** Ideal for open fires. Splits well, doesn't throw a lot of sparks and produces smoke with a strong rustic scent. Look for its distinctive annual rings. Best seasoned for a year.
- **Beech.** Splits easily when green, doesn't spark and produces flames that are a rich, bright yellow. A traditional choice for wood-burning ovens. Creates good embers.
- **Elm.** Tough, hard and cross-grained; must be split green. Its sharp smoke is fruity and fragrant.
- **Maple.** Dense and burns long and steady. Some varieties, such as field maple, can be difficult to split.
- **Black locust.** Hard, heavy and straight-grained. Easy to split. Has a perfumed smoke.
- **Hickory.** Tough, heavy, hard to split. Very high heat output and distinctive, aromatic smoke. A traditional favorite for grilling and smoking food.

SECOND CHOICE

- **Ash.** Pale wood, splits easily, with a gentle scent and very little smoke. Can be burned relatively green.
- **Birch.** Burns cleanly and sparks lightly with a lovely fragrance when burned. Signature bark can be used for kindling and is full of aromatic oils.
- **Walnut.** Dries slowly and splits fairly well. Average firewood with sharp, spicy smoke.
- **Cedar.** Dense for a softwood, burns well, richly scented.

THIRD CHOICE

- **Poplar.** Best for woodstoves and kindling; smoke can be bitter.
- **Pine.** Good softwood fuel but prone to throwing sparks and sooty smoke.
- **Spruce.** Poor firewood, slow-drying and dull-burning; little to no scent.

Just How Much Is a Cord, Anyway?

The official measurement of firewood is the standard, or full, cord, which is 4 by 4 by 8 feet (1.2x1.2x2.4 m). This works out to 128 cubic feet (3.6 m³) of evenly stacked wood, including bark and air space (see Fig. 2.1). But because 4-foot (1.2 m) pieces of wood are seldom used for wood burning, dealers rarely sell the full measure, instead offering a bewildering array of cords: face cords, stove cords, furnace cords, even pickup cords, each containing a different amount of wood. Most of these are 4 feet (1.2 m) high by 8 feet (2.4 m) long by whatever length the pieces of wood happen to be (see Fig. 2.2), the most common being around 16 inches (40 cm).

Because a winter's supply of firewood can easily run into thousands of dollars, it pays to shop around and compare the prices of several dealers. To avoid confusion and (sadly) price gouging, take a tape measure along and measure the length of an average piece of firewood being sold. If a dealer does not price the wood by the standard full cord, you can use this length to convert the price to this basic unit. Here are some examples:

A) Maple Grove Firewood sells a face cord for $95. Measuring the pile, you find it is 4 feet high by 8 feet long (1.2x2.4 m), with an average piece length of 16 inches (40 cm). Divide the depth of a full cord (48 inches/120 cm) by this length, then multiply by the dealer's price:

48 ÷ 16 = 3 x $95 = $285

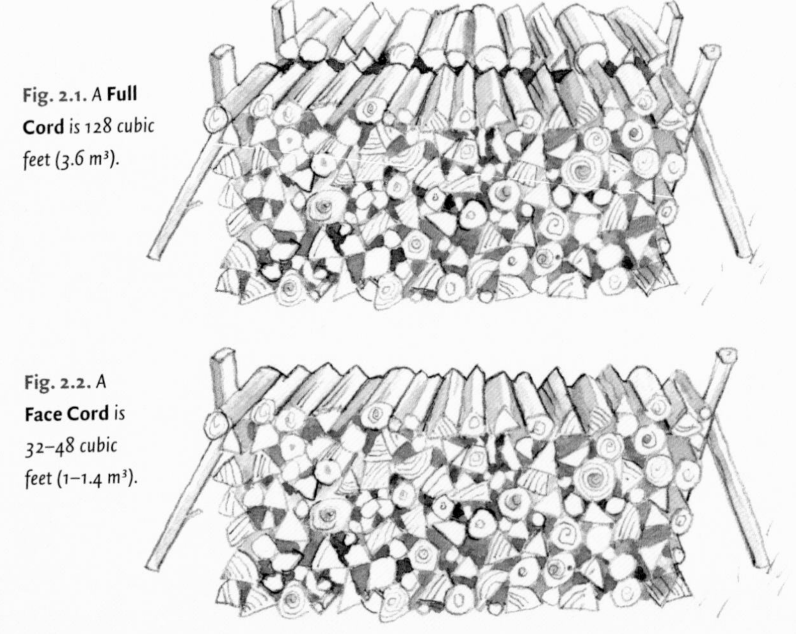

Fig. 2.1. A **Full Cord** is 128 cubic feet (3.6 m³).

Fig. 2.2. A **Face Cord** is 32–48 cubic feet (1–1.4 m³).

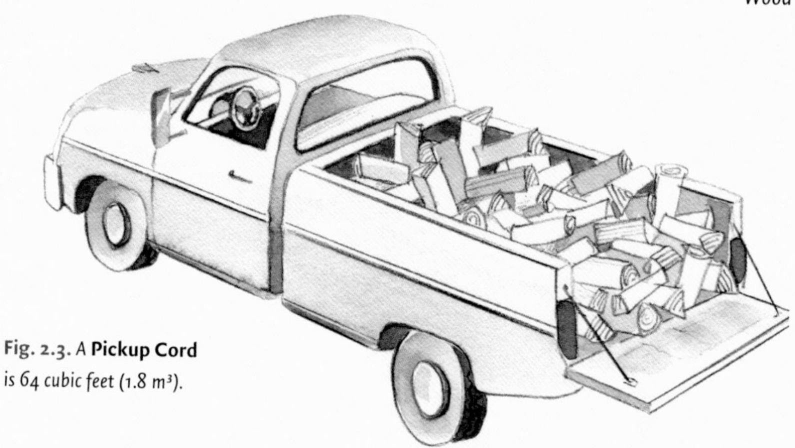

Fig. 2.3. A **Pickup Cord** is 64 cubic feet (1.8 m³).

Maple Grove Firewood thus sells firewood for $285 per cord. One full cord would require three of Maple Grove's face cords and cost a total of $855.

B) Down the road, Bud offers what he calls a "stove cord" for $50. The average piece length of the pile in Bud's woodlot measures 12 inches (30 cm). The calculation is:

48 ÷ 12 = 4 x $50 = $200

Bud sells firewood, then, for $200 per cord. You would need four of Bud's stove cords to equal one full cord, costing a total of $800.

C) Ironwood Home Heating sells a 4-by-8-by-1½-foot (1.2x2.4x0.5 m) furnace cord for $100. The calculation is:

48 ÷ 18 = 2.67 x $100 = $267

Ironwood Home Heating sells firewood for $267 per cord. Since it would take 2.67 cords of Ironwood's firewood to make one full cord, the total cost would be $712.89.

Therefore, Ironwood Home Heating has the best deal on firewood per full cord.

As a full cord is a large amount of wood, you will likely not be taking this home yourself—a cord of dried wood can exceed two tons in weight (three tons or more if it's green). Avoid buying firewood in delivery units that cannot be converted to a full cord. While common, station wagon loads and pickup loads (see Fig. 2.3) are imprecise measures, difficult to compare and usually conceal a higher price per cord. Beware of dealers who promise to deliver a cord of wood in a Ford F-150. A half-ton pickup with a standard bed can hold only half a full cord of firewood when loaded to the top of the bed.

Lastly, a neatly stacked cord has less air space and therefore contains more wood. The amount of solid wood in a full cord varies depending on the size of the pieces but usually averages around 85 cubic feet (2.4 m³). A randomly piled stack of wood contains more air and less wood. (Crooked, small-diameter, knotty or branchy pieces also reduce the amount of wood in a cord.) If you want to buy your cords tidily stacked, bear in mind that some dealers charge extra to stack the wood.

home, if heated exclusively with wood, burns through two to four full cords. If you're using a space heater for part of the heating load, you may require only one or two full cords, while reducing conventional fuel use by more than 50 percent. Considerably more wood is required in very cold areas, in large, leaky houses or where softwoods only are burned.

How much can you expect to pay for firewood? That, of course, depends on local supply and demand, the scarcity of the species, whether the wood is green or seasoned, whether it has been split before delivery and how the wood is sold (see "Just How Much Is a Cord, Anyway?" on p. 34).

Seasoning Wood

Wood is essentially a mass of tiny long tubes, or cell cavities, that run the length of the tree. Moisture exists both as "free water" in these cavities and as molecular water that is locked in the cell walls. When a tree is felled, the slow process of drying begins, and the free water is the first to evaporate. Once the free water evaporates, the moisture content of the wood is around 30 percent. This is called the "fiber saturation point." After this, water begins to leave the cell walls, and the wood starts to shrink and crack.

For optimal burning, firewood should be dried, or "seasoned," until its moisture content is less than 20 percent. Firewood with a moisture content higher than that may eventually burn, but it is devilishly hard to light and just as hard to keep burning. Also, your new high-efficiency wood-burning stove or furnace is guaranteed to perform sluggishly as it struggles to burn freshly split, or "green," firewood—much of the heat and energy content produced are wasted in drying the wood's excess moisture. Just as important, the stove does not burn the tars and creosote in the smoke produced by the fire, and they end up lining the inside of your flue pipes and chimney (see "Creosote: The Black Plague of Chimneys" on p. 88). They also blacken the glass windows of your wood-burning appliances and produce a lot of blue-gray smoke, fouling your house and annoying your neighbors.

Seasoning wood has another important but less obvious benefit—when wood is properly cut and stacked right away, mold has less

Fig. 2.4. Testing wood with a moisture meter.

opportunity to establish itself. Throwing unseasoned firewood into a pile allows mold to spread throughout the logs, mold that you unwittingly release into your home's environment when you bring the firewood inside throughout the heating season.

You can buy already seasoned firewood or buy it green and season it yourself. How can you tell whether wood is properly seasoned? It's possible to test the wood with a moisture meter (see Fig. 2.4 above), which measures resistance to a small current and converts it into a moisture-content reading, yet this reading can vary widely from one area of the log to another. With a little practice, however, you can use the following tips to judge accurately for yourself whether your wood is dry. Use as many as you can for the best results.

Radial checking. Look for cracks and checks in the end grains that radiate out from the heartwood to the sapwood. These appear before the wood is totally seasoned, so your testing should not stop here.

Color. Wood fades and darkens as it seasons, changing from white or cream to yellow or gray. Different species have different colors and shades, but it's safe to say a stack of bright, freshly colored wood is far from seasoned.

Smell. Split a piece and sniff; if the exposed, fresh-cut surface has a pleasant, sappy aroma (or if it feels damp and cool), it's too wet to burn.

Loose bark. As wood dries, the bark slowly begins to separate from the wood and eventually falls away. If the bark is still attached to the wood, peel it back with a sharp knife and check the cambium. If the cambium is green, so is the wood. A cord of seasoned wood should have more wood without bark than bark-covered wood.

Listen. Bang two pieces of wood together. Dry wood sounds hollow; wet wood sounds dull.

Lift. Seasoned wood weighs much less than green wood of the same species.

Trial by fire. If in doubt, burn some! Dry firewood ignites and burns easily; wet wood is tough to light and hisses in the fire.

There *are* advantages to buying firewood green, provided you have the room to store and season it for a year. For one, you'll be absolutely sure the wood is seasoned, and it will cost a lot less, anywhere from $15 to $50 less per cord than seasoned wood. Also, seasoned wood may be in short supply in some areas, so you may not have a choice.

Firewood can take a very long time to properly season. Exactly how long is a matter of ongoing debate in wood-burning circles. The traditional rule of thumb is to season firewood for at least six months before the heating season; some hardwoods require at least one to two full years. The truth lies somewhere in the middle and depends on piece size, tree species and local climate.

The protective bark on a log helps prevent the interior moisture from evaporating, so firewood begins to dry significantly only after it is cut and split. By splitting the wood into smaller pieces, you create a greater surface area, and the greater the total surface area, the lower the overall density, which means the wood dries and seasons at a faster rate. Trees with a dense wood structure, such as oak and elm, season much more slowly than do ash and birch. "Diffuse porous" species, such as maple, birch and poplar, season more quickly than do "ring porous" species, such as oak and ash. Conifers have an entirely different cell structure than deciduous trees and take longer to dry, so they are best split into small pieces. Trees felled in spring when the sap is "up" also have a higher moisture content. Finally, if you live in a damp maritime climate, seasoning times may be longer.

As with a really good meal, seasoning makes all the difference in a quality fire. In the end, with the exception of truly dense hardwoods, such as oak, and large-split softwoods, most household firewood bought in the spring can be seasoned enough for burning by winter. It's not the type of tree but, rather, the seasoning that makes or breaks your fuel supply.

If you plan to season firewood yourself, here are five simple guidelines to follow:

Cut to length. Cut firewood to the right length for your stove, fireplace or furnace. This is usually about 3 inches (7.5 cm) shorter than the width or length of your firebox, depending on how you load the wood. Shorter is always better than longer.

Split to the right size. Wood should be split to the proper dimension for your wood-burning appliance. For most efficient woodstoves, that is no more than 6 inches (15 cm) across. A range of dimensions from 3 to 6 inches (7.5–15 cm) for woodstoves and slightly larger for furnaces is best.

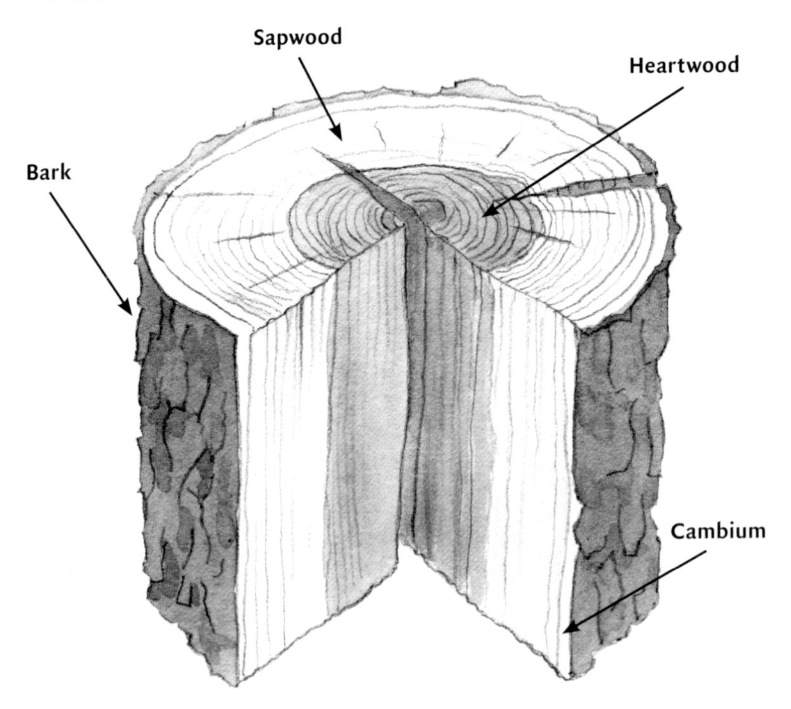

Fig. 2.5. Stump cross section

Stack and expose. To season firewood properly, stack it in a place where the sun can warm it and the wind can blow through it. A single row exposed to the sun and prevailing winds is best—as the sun heats and evaporates the water from the wood, the wind whisks it away (see "Stacking and Storing Wood" on p. 48).

Season for a season. The key to seasoning lies in the word itself: Most firewood properly split and stacked takes at least a season to dry properly. For many of us, that is about six months. If you stack your wood in early spring, it should be ready to be put away for winter use by October. Hardwood may take longer depending on the species, the local climate and how green it is when you buy it.

Don't cover it up. Covering your drying woodpile may keep the moisture off your firewood if you live in a rainy climate, but it can also hinder the sun's drying and lead to additional chores, like chasing tarps or plastic sheets blown away by the wind.

Splitting Wood

Unless you enjoy the challenge of lighting a log, you need to do a little splitting to reduce your newly purchased firewood to a manageable size for burning. Of course, you could have the supplier do this for you at a cost. You could also save your money and do your own splitting. Make no mistake—splitting a cord of wood to size is an enormous amount of work. Even so, in the end, it can be an immensely satisfying chore.

When considering where to do your splitting, keep three key things in mind. Safety first. You need a lot of space to swing an axe. Choose a splitting area that is well away from structures and, most important, people. Make sure your field of vision is uncompromised so that no one, including children, can approach you unobserved.

Second: After you split a cord or so, the area around the chopping block may look like a trench on the front line—muddy, trodden and littered with wood chips. Pick a spot where that kind of mess won't bother you.

Third: Avoid unnecessary damage. Make sure there are no drains, pipes or other pieces of infrastructure buried under your chopping

Freeing a Stuck Axe

From time to time, you will end up with your axe stuck in a log and find yourself feeling like King Arthur trying to free Excalibur from the stone. To unstick it, try these four basic techniques, in successive order:

1. Tap the end of the axe handle sharply downward with your palm, though not hard enough to damage your hand. Stand clear and to the side of the arc of the freed axe in case it jumps back out.
2. Carefully lift the axe and log (if they aren't too heavy), turn them both over and bring the axe down hard on the chopping block, letting the weight of the log do the work.
3. If your axe is embedded in a particularly large log, tap the side of the axe head gently with a lump hammer, first on one side, then on the other. This might free it.
4. As a last resort, use a hammer and wedge. Set the wedge in the log in line with the stuck axe. Carefully hammer the wedge in; this should widen the split and free the axe.

After your axe is free, always inspect its head and handle for damage and make sure the head has not loosened.

area. Repeated blows reverberate beneath the ground, and the last thing you want is to disturb or destroy a water line.

A log splits easily when it is caught between two things: a descending axe and a stationary block, so put some thought into your chopping surface. (Without a block, there's nothing to absorb the blow but the ground.) If you can lay your hands on one, an old tree stump is always a popular choice, particularly dense species such as oak. The block height should be about 6 inches (15 cm) below the level of your knee. Any higher, and it may topple over, and after you've spent an hour or so lifting and placing a succession of pieces of heavy, unsplit wood, it will take a toll on your back.

Dress for the occasion by choosing clothes that are old, comfortable, warm and loose-fitting, with steel-toed work boots for safety. Gloves are not recommended, as they could prevent you from getting a tight grip on the axe. You don't want to risk its flying out of your hands. Eye protection is also vital, as wood chips and even bits of metal will be swirling around.

When choosing a splitting tool, consider a maul rather than an axe. An axe has a heavy, flared head and a thin, sharp blade. A splitting maul has a steep, sloped wedge and a blunt end, which increases outward pressure on the wood and is less likely to get stuck than an axe. An axe is good for chopping smaller pieces of firewood; the maul

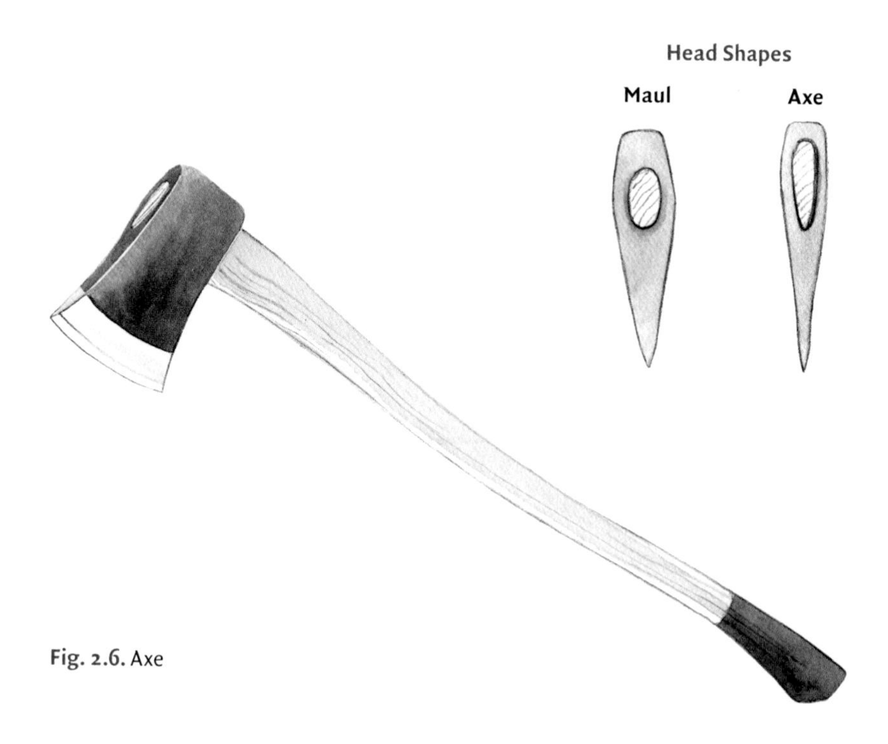

Fig. 2.6. Axe

is better for getting those bigger logs cut to chopping size. While a 4-pound (1.8 kg) axe can handle most logs, you will get more chopping mileage out of a 6-pound (2.7 kg) splitting maul. Steer clear of the 8-pound (3.6 kg) or 10-pound (4.5 kg) models unless you are exceptionally strong; a lighter maul can be swung much faster and with greater force. Choose something you would be comfortable hefting and using for over an hour. Never use a double-bladed axe to chop wood; the chance of getting hit with a blade when extracting a stuck axe (see "Freeing a Stuck Axe" on p. 42) is not a risk you want to take.

Splitting Wood Step-by-Step

1. Place the log on its end in the center of your chopping block, with any knots or crotches angling downward. Stand facing the upright log, feet in line with your shoulders. If you can, position yourself slightly uphill. This will bring your weight and height to bear in maximizing the effectiveness of the blow.

2. Study the log to be split for cracks or signs of weakness, and choose these as your target. Avoid splitting through knots or branches.

3. Aim to make contact near the edge of the log, NOT the center. That way, the maul strikes at a 90-degree angle to the growth rings, which is where the log is weaker. Remember, it may take several blows to fully split a log.

4. Measure your distance from the piece of wood by placing the maul where you wish to strike with arms fully extended, and then step back a half step. That way, you'll have to lean forward a little, which

Fig. 2.7. Chopping block

adds power to your swing. With elbows comfortably bent, hold the maul horizontally near waist level, one hand at the base of the handle and palm facing you, the other hand at the neck, thumb next to the maul head and palm facing away from you.

5. Flex your knees, and bend slightly at the waist. Quickly raise the maul overhead, arms high, and straighten your back and knees. During your upswing, let the hand next to the maul head slide down the handle to meet your other hand.

6. With no delay, swing down forcefully. Your swing should come over your head, not your shoulder. Focus on the point of intended impact. Bend your waist and knees, putting all of your body into the downswing. Do not let your sight wander from the striking point.

Fig. 2.8. Proper stance and distances from chopping block for splitting wood.

7. Knotty or twisted-grain wood may need a steel splitting wedge and a sledgehammer to pry open. Set the wedge in the widest point of the crack, and drive it in with the sledgehammer as you would a large nail. The technique involved in swinging remains the same.

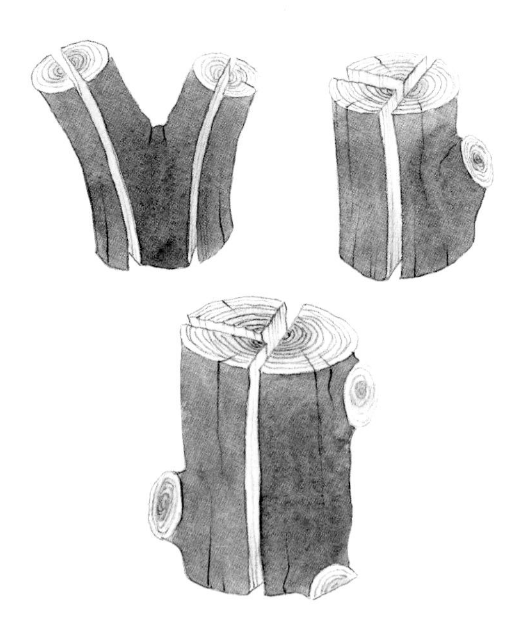

Fig. 2.9. Planning difficult splits

Never drive a wedge so far into a piece of wood that you can't knock it back out, and always wear eye protection.

When splitting wood, keep in mind the type of firewood you want. To build good fires, you need a range of sizes, from 1- to 2-inch (2.5–5 cm) pieces of kindling to small pieces of wood for short, hot fires or small fires in mild weather and large pieces for longer firing cycles. A selection of pieces from 3 to 6 inches (7.5–15 cm) in diameter for woodstoves and an inch or so larger for fireplaces and wood-burning furnaces should serve you well. Remember: Big, unsplit pieces make lousy firewood.

To create kindling, use logs with a straight grain and no knots. A simple way to make kindling is to slice off thin slabs of a log and split these shingles into sticks about 1 inch (2.5 cm) wide with a small axe. Large logs can be split into rectangular pieces suitable for stacking. Small-diameter logs can also be split into wedges or pie shapes that will not roll out of an open fireplace.

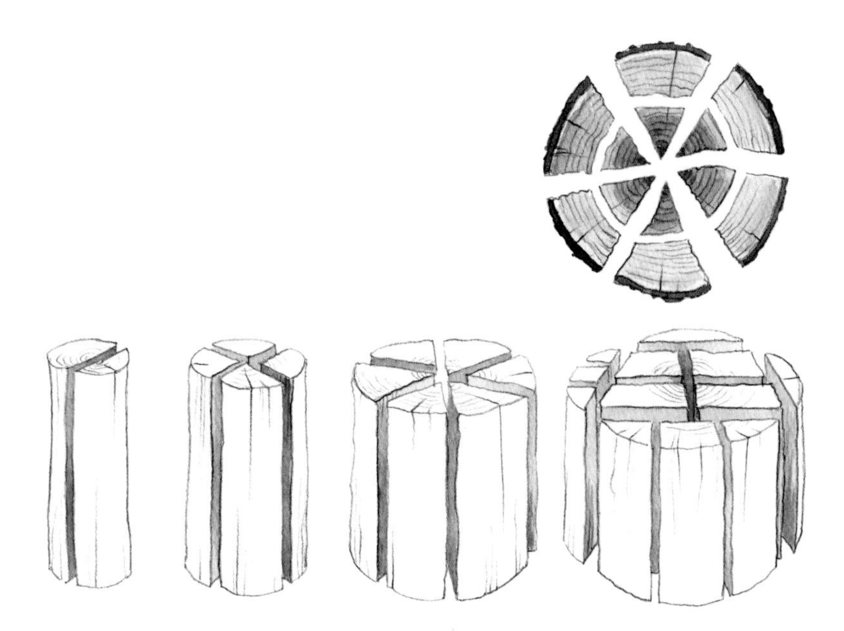

Fig. 2.10. Splitting patterns

Wood-Splitting Tips

- Hardwood logs split most easily when they are green, so if possible, split them before they are thoroughly seasoned. Conifers are the opposite—you will quickly get covered with sticky sap and resin if you split them before they are dry.
- Wood splits easily when frozen. The colder, the better!
- It's easier to split a piece of wood at the end farthest from the knots.
- A slightly sloping surface to your chopping block (less than 10 degrees) allows you to rotate logs on their end until they stand up straight.
- Learn to strike within a quarter of an inch (6 mm) of your intended spot. If you strike accurately, the wood will split cleanly and you won't be hit with flying bits.
- If you use a sledgehammer and wedge, keep an old axe handy to clean up the strands of stringy wood.
- When you are tired, stop splitting. The risk of injury is far greater when you are fatigued.

Stacking and Storing Wood

It is often said that a long, straight row of firewood standing in the yard in springtime is like money in the bank. As it dries in the summer wind and sunshine, you're collecting interest that will pay off when winter arrives. But all the saved-up summer heat that you're hoping to unleash in your fireplace, stove or furnace could be wasted if you neglect the simple chore of stacking and protecting your fuel.

Stacking firewood properly allows it to season and prevents mold. Well-stacked wood is also more pleasing to the eye than an unkempt pile. Water is the chief element against which you need to guard: Exposed to rain, seasoned firewood quickly becomes unseasoned as it regains the moisture it has lost. Wood piled under dripping eaves, for example, will never season.

There are three simple rules for good wood stacking:

1. **Allow air circulation.** Leave the sides of your stacked woodpile uncovered, and avoid stacking your firewood as though it were a tight-fitting jigsaw puzzle. Stack it in long, narrow rows that are one stick wide. Make separate rows if you have to, but be sure to leave some space between rows for the sun and wind to penetrate the stacks. Stacking them north-south allows you to take advantage of east-west winds. Make the stacks no higher than 4 feet (1.2 m) to avoid the risk of their toppling over. Look for the best place to dry your laundry; if space permits, this would be a good place to store your wood.

2. **Shelter it from the rain.** You may need to cover the top of your stacks with a tarp, but if you do, leave some room between the tarp and the top of the wood to allow air to circulate. And make sure the tarp extends beyond the woodpile so that any precipitation runs off. Do not cover the sides of the stacks, as this will hamper drying. If you live in a fairly dry climate and don't want to cover up your wood, half-round logs placed on top, bark side up, will shed rain more easily and keep the pile dry.

Fig. 2.11. Covered wood-storage area

3. **Keep it off the ground.** Firewood needs to be protected from water from below as well as above. Mold and rot can set in quickly after just a few days on exposed ground, and seepage could affect as much as 10 percent of a typical stack. Wooden pallets, poles or lumber rails are all great ways to keep your wood high and dry.

To avoid having to rebuild your woodpile after a strong frost heave undoes all your hard work, you'll need to support the ends of each woodpile. Depending on the length of the stack, the sides might need support as well. Drive stakes or thin trunks into the ground along the length of the pile to support the sides. To support the ends, drive supporting stakes into the ground, attach the end pieces to pallets or build retaining "walls" by cutting logs into rectangular pieces and stacking them, log-cabin-style, at either end.

Finally, give some thought to where you locate your stacked woodpile. During the winter, you're sure to want it as close to the house as possible. Termites, ants, earwigs, hornets and mice would all agree. Pest-control experts, however, suggest that the farther it is from your house, the better. Around 150 feet (45 m) away is a good compromise.

When your firewood has dried and seasoned to below 20 percent moisture in the summer sun and breeze, it's a good idea to move it to a dry storage area for the winter, if possible. This area should be dry, fully sheltered from rain and snow and close to—but not inside—the house.

A good wood-storage area should feature all of the following characteristics:

- It is well ventilated, has a sound roof and offers protection from wind and rain.
- It holds at least a year's supply of wood for your home.
- There is sufficient room to store different categories of wood, such as hardwoods for extended firing cycles, softwoods for short, hot fires in spring and fall and separate piles for cooking and heating.
- It includes a place for splitting and chopping kindling.
- It blends in with the surroundings.

It's generally not a good idea to store large amounts of wood inside the house unless you want to invite mold, dirt and insects in as well. However, a small amount of wood—two to three days' worth—stored inside a wood box away from the fire will prevent frozen logs from shocking and dampening your fire.

You can eliminate the mess involved in moving wood with a little planning. By the time your firewood is ready to be brought into the house from the woodshed, most of the wood chips, bark and bugs beneath it will be left behind. Invest in a good-quality wood carrier—a tough canvas bag with handles—to make hauling wood from the shed to the wood box easier and neater.

Chapter 3

Appliances

WOOD-BURNING APPLIANCES COME IN a bewildering array of models and systems, and determining which one will be a good fit for your house and your home-heating objectives can be a challenge. Do you simply want to relax in front of the occasional fire? Do you want to trim your home-heating bills? Or do you want to live completely off the grid?

Choosing a Wood-Heating Strategy

When shopping for a wood-burning appliance or system, consider the following wood-heating strategies: wood as the *sole* heat source, wood as a *primary* heat source, wood as a *supplementary* heat source, wood as a *recreational* heat source and wood as an *emergency* heat source.

Very few homes fit into the first category. If your goal is to rely on wood as your sole heat source, your house must be in a rural setting near a plentiful source of firewood and, ideally, the building should have a relatively open concept that allows for central heating. Be prepared to be home 12 hours a day, because tending a fire in a furnace or stove is a full-time job.

If wood is your primary source of heat, you may not have to tend your stove or furnace 24/7, but you'll be burning nearly as much wood.

Table 3.1

Wood-Burning Appliance Particulate-Matter Emission Limits		
Type of Appliance	U.S. EPA Standard (2012–2016)	CSA B415.10 (2012–2015)
Catalytic wood-burning devices	NSPS† 2013: 2.5 g/h	2.5 g/h
Non-catalytic wood-burning devices	NSPS 2013: 4.5 g/h	4.5 g/h
Low-mass factory-built fireplaces	EPA Phase 2: 5.1 g/h	Currently no limit
Site-built masonry heaters	Currently no limit; NSPS 2014 (proposed): 2.0 g/h	Currently no limit
Site-built masonry fireplaces	Currently no limit; certification of masons proposed	Currently no limit
Decorative factory-built fireplaces	Currently no limit	Currently no limit
Indoor boilers and furnaces	NSPS 2014: 0.32 lb/MMBtu* heat output	0.4 g/MJ
Outdoor boilers	NSPS 2013: 0.32 lb/MMBtu heat output	0.13 g/MJ

†New Source Performance Standards under the U.S. Clean Air Act
* MMBtu equals one million BTUs

Source: *Code of Practice for Residential Wood Burning Appliances*, Canadian Council of Ministers of the Environment, 2012

Your house, location and lifestyle must all be adapted to accommodate those realities. Fire is a hard taskmaster.

If wood provides you with a supplementary source of heat, you are leveraging wood burning to save on utility bills and using space heaters or fireplaces for cold snaps. This option is available to many urban and rural dwellers.

Lastly, the option of using wood as a recreational or an emergency source of heat is available to everyone who enjoys a good fire—and can be a godsend when power lines go down.

Each of these strategies calls for a different wood-burning appliance. These appliances fall into three basic categories: **space heaters** (woodstoves, cookstoves, pellet stoves), **fireplaces** (including fireplace inserts and masonry heaters) and **central-heating systems** (furnaces and water boilers). The most suitable choice depends on your heating needs; the construction, location and layout of your home; the local climate; and how much time and energy you're willing to devote to the wood-burning lifestyle.

Space Heaters

A space heater is meant to heat a space directly, unlike a central-heating system, which distributes heat to the entire house through a system of ducts or pipes. The most common space heater is a wood-burning stove, but cookstoves and pellet stoves are becoming increasingly popular.

Once upon a time, when homes were poorly insulated and drafty, a space heater could be expected to heat only the room in which it was installed. In today's energy-efficient houses, which retain heat more effectively, a single space heater can heat far more than its immediate space, provided the installation and the means of directing heat are effective.

If you plan on meeting most of your home-heating needs with a space heater, be sure to locate the heater in the area where your family spends the majority of its time (see "Installing a Wood-Burning Appliance" on p. 93).

Wood-Burning Stove

Surveys show that between 60 and 80 percent of homeowners who produce at least some of their heat with wood do so with a wood-burning stove, probably because a woodstove is the most common, flexible and inexpensive option for heating a small space. A happy compromise between the unsightly, utilitarian furnace in the basement (which heats the whole house) and the aesthetically pleasing fireplace (which heats only the living area), a woodstove can be located in almost any room where there is enough space for both the stove itself and a properly routed chimney. Because modern houses conserve

Certified Wood-Burning Appliances

The internal design of wood-burning stoves has changed entirely since the early 1990s, largely as a result of the U.S. Environmental Protection Agency's *Standards of Performance for New Residential Wood Heaters*, Section 60-532 of the 1988 *Clean Air Act*. A decade later, in 2000, the Canadian Standards Association developed the *Performance Testing of Solid-Fuel-Burning Heating Appliances* (CSA B415), based on the EPA regulations. Both standards require independent testing by an accredited laboratory and specify tests for measuring the emissions, heat output and efficiency.

While older, uncertified stoves release 15 to 30 grams of particulate matter—smoke—per hour, the EPA regulations set the mandatory smoke-emission limit for woodstoves at 7.5 grams per hour (g/h) for non-catalytic stoves and 4.1 g/h for catalytic stoves (see "Cats & Non-Cats: Two Different Breeds of Stove" on p. 62). In 2013, the EPA raised these limits to 4.5 g/h and 2.5 g/h, respectively. The Canadian standard was updated in 2010 (CSA B415.10) and is currently equivalent to the U.S. EPA standard (see Table 3.1 on p. 54).

Today, all woodstove and fireplace inserts (and some factory-built fireplaces sold in the United States) must meet these limits. (In Canada, EPA-/CSA-certified wood heaters are mandatory in British Columbia, Quebec, Nova Scotia and Newfoundland and Labrador.) Certain municipalities may also have bylaws that require the installation of EPA-/CSA-certified wood heaters, even though there are no provincial regulations on the books.

Over the years, woodstove manufacturers have worked within these regulations to make vast strides in combustion technologies. On average, advanced EPA-/CSA-certified stoves are about 30 percent more efficient than old box or potbellied woodstoves and almost all the currently available central wood-burning furnaces and boilers. They produce almost no smoke and minimal ash and require less firewood. Advanced stoves also produce about 90 percent less particulate matter and 90 percent less creosote than older stoves. The result? No nuisance smoke, fewer chimney cleanings and a reduced risk of chimney fires. The extra cost of advanced technology is only around US$200 per stove, a cost that is easily recovered after a few seasons of heating with wood. When comparing stoves, be sure to look for EPA and CSA certification labels on the back.

energy more effectively than older houses and require less heat to stay warm, it's possible to heat an average-sized modern home with a single wood-burning stove, provided your home is open and the stove is centrally located. The downside to these stoves is threefold: Operating woodstoves can be a messy proposition, they dry the air in your house, and their fires require a lot of tending.

Woodstoves were once seen as the bumbling, scruffy relations of the home-heating family, perpetually leaking heat and depositing dangerous amounts of creosote in chimneys. That all changed at the end of the 1980s, when new North American regulations on emissions of wood-burning appliances were developed (see "Certified Wood-Burning Appliances" on p. 56). Today's advanced-technology wood-burning stoves are highly efficient, nearly smokeless appliances that can heat not only a small space but, under specific circumstances, an entire house.

When shopping for a wood-burning stove, think about whether catalytic or non-catalytic combustion (see "Cats & Non-Cats: Two Different Breeds of Stove" on p. 62) or radiant or convection heat makes better sense in your situation. A radiant stove—such as cast-iron stoves and those with heavy steel-plate surfaces—sends its heat out in all directions and is effective in a relatively open area, where the radiant warmth is dispersed to the walls, floors, ceilings, furniture and people that face the appliance. As appealing as this might sound, however, there's a downside: A radiant stove in a small room can throw off enough heat to make those people feel uncomfortably warm.

A convection stove heats air that flows between the stove body and a sheet-metal shield or casing. The shielded surfaces don't get as hot as radiant surfaces do, which makes this stove a better choice for small rooms, where people might inadvertently come into contact with the stove. Many modern stoves feature a blend of both characteristics: The back and sides may have convection shields behind which air flows, with radiant front and top surfaces.

Woodstoves are not designed to be central-heating appliances, which means they should be located where heat is needed. Indeed, choosing the right location for your woodstove may be the most

Fig. 3.1. Woodstove

important heating decision you make. Install it in the part of the house you want to be the warmest. This is usually the main floor area where kitchen, living and dining rooms are located and where families typically spend most of their time. If your home design is open concept with few partitions, you may be able to heat the whole building with a woodstove that is the right size and efficiency.

A basement is not a good location for space heating, unless the basement is the space you want to heat. Yes, heat rises, but it does so too slowly and inefficiently to provide any comfort on the upper floors. To keep the main floor comfortably warm, you inevitably end up overheating the basement. Unfinished basements are even worse because so much of the heat is absorbed by the walls and lost to the outside. If you want a wood burner in the basement that will heat the upstairs rooms, consider a wood-burning furnace or boiler (see "Central-Heating Systems" on p. 72).

Choosing the appropriately sized woodstove for your space is also critical. A stove that is too large for the room quickly overheats the space, while a stove that is undersized can be damaged as a result of overfiring (i.e., building a fire that generates too much heat for the size of the firebox in order to maintain a constant room temperature). Selecting the right stove to match your needs can be tricky—performance is often related to the size of the firebox rather than the overall size of the stove itself. If in doubt, visit a certified wood-heat retailer with a floor plan of your house in hand and ask for expert advice.

Pros

- Price: A wood-burning stove is the least expensive and most flexible of space heaters.
- Highly efficient, advanced-technology models boast reduced emissions and less smoke.

Cons

- A wood-burning stove heats only the surrounding area and is not suitable as a main heating system for large homes with many small rooms.
- It tends to dry the air in the house.
- It cannot be installed in the basement.

Cookstove

Wood is a fuel that can warm you many times over: when you cut it, when you split and stack it, when you burn it in the fireplace and, finally, when you prepare a home-cooked meal over it in a wood-burning stove. Just ask anyone who has had the pleasure—nothing tastes like a pie, a pot roast or coffee that has been prepared over a wood-fired stove.

Originally developed in the 1800s, the cast-iron cookstove quickly became a staple in many homes for its ability to both warm the house and bake the daily bread. Many of the wood-burning cookstoves produced today are replicas of these original models with modern updates incorporated.

The wood-fired stove is designed for stovetop cooking and oven baking, and it may also have a warming closet and a hot water jacket in which to boil wash water. Indeed, a good cookstove can be a multitasking wonder—cooking appliance, high-output room heater,

Fig. 3.2. Cookstove

attractive fire-viewing woodstove and, with the water jacket, a source of hot water.

While a cookstove can be used to heat limited areas, space heating is not its main function. Without a thermostat to guide you, mastering the idiosyncrasies of temperature in a wood-fired oven takes time and experience. Knowing exactly how much wood to burn to achieve and maintain a desired temperature for baking comes only with much trial and error (see "Cooking on a Wood-Burning Woodstove" on p. 128). Because there is no advanced-technology wood-burning cookstove, a cookstove's smoke emissions are higher and its efficiency is lower than in advanced woodstoves. Lastly, choices are limited: As demand for wood-burning appliances declined over the past 50 years, many North American stove manufacturers went out of business or dropped specialized models from their product lines. What often remains are retro and/or high-end models, and they are priced accordingly.

Cats & Non-Cats: Two Different Breeds of Stove

To meet CSA and EPA guidelines on efficiency and smoke emissions, manufacturers of wood-burning stoves have developed technologically advanced wood-burning appliances that employ one of two different methods of combustion: catalytic and non-catalytic. Both methods are effective and can hold a fire overnight, but there are subtle performance differences.

Catalytic Combustion

During catalytic combustion, the smoke produced by the fire is passed through a coated ceramic catalyst inside the stove where the gases and particles ignite and burn. The platinum or palladium coating on the ceramic honeycomb lowers the ignition temperature of the smoke as it passes through. This allows the catalytic appliance to operate at a low firing rate, producing a long, even heat output while still burning cleanly. Because the catalyst restricts gas flow through the appliance, the appliance always includes a bypass damper that is opened when the unit is loaded. When a hot fire is established and the catalytic converter reaches a light-off temperature of 500 to 700 degrees F (260°C-370°C), the damper closes, forcing the smoke through the combustor for an extended clean burn. The catalyst itself then burns much of the smoke before it reaches the flue, reducing creosote by up to 90 percent. While the catalyst degrades over time and must be replaced (it can last for more than six seasons if the stove is used properly), the most pleasant feature of the catalytic stove is that it produces more heat per cord of wood.

Non-Catalytic Combustion

A non-catalytic high-tech stove relies on secondary combustion instead of a catalyst to achieve a clean burn. It does this by guiding smoke from the burning wood to hot spots in the firebox and introducing it to a fresh supply of combustion air, with the resulting high temperatures burning off the smoke. Three internal features create the temperatures required for complete combustion: (1) firebox insulation; (2) baffle plates that reflect heat back into the firebox and create a longer, hotter gas flow; and (3) a heated secondary air supply usually fed to the fire through ducts with small holes above the fuel bed. Although top-of-the-line non-cats offer a beautiful fire and high performance, they cannot match the even heat output of a catalytic stove. The baffle and some other internal parts of a non-catalytic stove need replacement from time to time as they deteriorate with the high heat of efficient combustion. Because catalytic stoves are slightly more complicated to operate, non-catalytic combustion has become the dominant advanced technology used in wood-burning appliances.

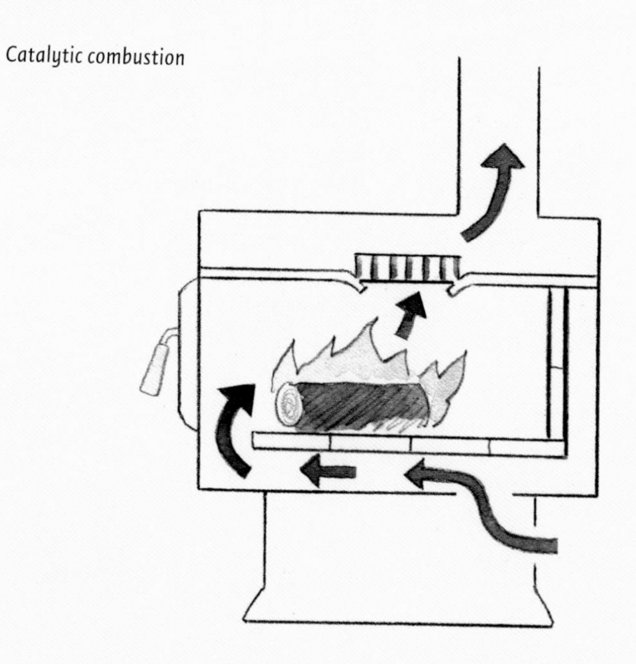

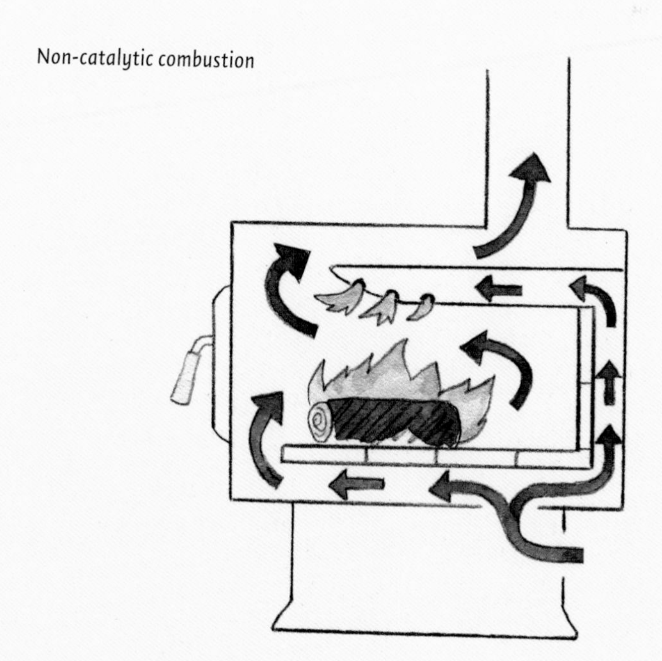

Fig. 3.3. Catalytic and non-catalytic woodstoves

Pros
- A cookstove can double as a space heater and a stove for cooking.
- It provides hot water for washing.
- Its old-style design can add rustic charm to your home.

Cons
- Price: Most high-end designs are two to three times the cost of an advanced-technology woodstove.
- A cookstove must be placed in the kitchen, which limits its space-heating function.
- Cooking and baking with a cookstove are a challenge and require getting a feel for how the stove burns.

Pellet Stove

Popular in urban areas and in areas where conventional wood burning is impractical, this stove is fueled by pellets of dried and ground waste wood, paper and other biomass materials that have been compressed into small cylinders about a quarter of an inch (6 mm) in diameter and 1 inch (2.5 cm) long.

The inside workings of a pellet stove are more complex than those of a conventional woodstove. Most important, it is electricity-dependent. A motorized auger transfers the pellet fuel from a hopper into a combustion chamber, an exhaust fan forces the gases into the venting system and draws in combustion air, and a circulating fan moves the heat into and around the room.

A pellet stove can hold at least 45 pounds (20 kg) of fuel and usually burns cleanly because the pellets are fed to the combustion chamber at an even rate. Many are thermostat-controlled. Pellets are far drier than seasoned wood (5 percent moisture content), which results in low emissions and high efficiency ratings similar to those of a catalytic stove. Because the draft is forced and the exhaust cool, it is possible (depending on the manufacturer's recommendations and local building codes) to vent a pellet stove to the outdoors with a double-walled flue pipe instead of a chimney.

The pellet stove is perhaps the easiest wood-burning appliance to install and use. However, wood pellets look like rabbit food, with

Fig. 3.4. Pellet stove

none of the aesthetic and romantic appeal of burning firewood, and this stove is effectively dead in the water when the power goes out.

Pros

- A pellet stove is a good choice for urban areas, for areas unsuited to conventional wood burning and for areas where firewood is unavailable.
- It has lower smoke emissions than a conventional wood-burning stove.
- Its heat output is steady and convenient, with an automatic operation (one hopper load can last 24 hours or more).
- It is easy to install; in some cases, it can be installed without a chimney.
- Pellets don't require splitting, stacking or seasoning and don't create firewood "mess."

Cons

- A pellet stove tends to cost more than a conventional wood-burning stove.
- Pellets can't be personally harvested like firewood and have no firewood "feel."
- The pellet stove is dependent on electricity for the auger, the forced-air intake and the internal fans, which can be noisy in some models.
- Pellets can be expensive. At around US$5.98 per 40-pound (18 kg) bag, the typical burn rate of two bags per day during winter for most homes can cost more than firewood in some areas.

Fireplaces

The hearth at the heart of many a home, the fireplace is a time-honored source of warmth and comfort. The fire that crackles and roars within bathes the room with a cozy, soothing radiance no other wood-burning appliance can achieve. A fireplace brings us as close as we can get to the sweet fragrance of wood smoke, dancing orange-yellow flames and the mesmerizing reds of hot, glowing embers. Outside of rural areas, a fireplace today is used mainly for fire viewing rather than heating. Yet although we may no longer need fire as a source of warmth or for domestic use, fireplaces continue to be built. Some are built on site during home construction from masonry materials like brick or stone; some are factory-built from steel.

Conventional Fireplace

As a home-heating alternative, a conventional masonry fireplace is a poor choice, as it famously removes more heat from the living area than it produces. It consumes huge amounts of air—much more than is needed for combustion—effectively cooling the fire and reducing draft, and its large flue allows most of the generated heat to be drawn up the chimney. (While other wood-burning appliances heat your home fairly quickly, they lack the thermal mass to store this heat. A conventional masonry fireplace retains and radiates this heat much longer.) A conventional fireplace also pollutes the indoor and outdoor air more than does an advanced-technology wood-burning appliance. This was not such an issue generations ago, when homes were roomier and draftier, but in today's smaller airtight homes, it can be a nuisance as well as a health hazard.

If you have your heart set on a traditional hearth and already have a conventional masonry fireplace that you wish to upgrade, simply installing a glass-door assembly and closing the doors when the fire is burning sharply reduces the fire's demand for combustion air and forces the average exhaust temperature way up. While this does not significantly improve energy efficiency, it does create a stronger, more stable draft and cuts down on nuisance smoke both indoors and out. You could also install a fireplace insert or a hearth-mount stove (see "Fireplace Insert" on p. 68). These units are effective heaters used by many North Americans to reduce home-heating costs. They offer an excellent view of the fire, thus preserving one of its original benefits.

If you plan to install a new fireplace, you can combine the beauty of a traditional hearth with the heating power of a modern woodstove by choosing an advanced-technology, factory-built fireplace. The same technologies used in woodstoves to meet regulated smoke-emission limits—gasketed glass doors, carefully designed combustion chambers and heat exchangers to return the heat to the room—are also found in these specialized units. Certified for low emissions, these appliances are equipped with internal baffles, firebox insulation and strategically placed combustion air inlets to produce a stable, clean-burning fire. You can trim the front of your fireplace with ceramic tile, brick or stone and add a decorative mantel. In most cases, an advanced-technology fireplace can be installed on a typical house floor without the need of a foundation or reinforcement. While this is a more sensible alternative to the conventional fireplace, note that its installation is somewhat complicated and is best left in the hands of trained professionals.

Lastly, if you haven't broken ground yet on a new home and plan to build a fireplace, consider including a cold-air return in the design. This is a channel that extends from the inside edge of the hearth to the outdoors, allowing the fireplace to draw its combustion air from outside the house rather than from the room you plan to heat.

Fig. 3.5. Fireplace

Pros

- A fireplace appeals to the senses and remains the most popular method of enjoying the romance of fire.
- A fireplace is already installed in some homes.
- Seasoned wood stacked and burned properly burns very cleanly and produces little creosote.
- Advanced-technology fireplace models are readily available.

Cons

- A fireplace is not suited for heating the entire home.
- An open fireplace tends to be smoky and messy.
- Fires require more building, stoking and tending than they do in a woodstove.
- Fireplace chimneys require annual cleaning.

Fireplace Insert

The history of the fireplace insert goes all the way back to Benjamin Franklin's 1741 eponymous Franklin stove, circa 1741, a metal box with hollow baffles and an "inverted siphon" built to fit into an existing brick fireplace. Today's fireplace inserts are essentially high-performance woodstoves designed to convert existing inefficient, leaky masonry fireplaces into effective heating systems. An outer steel shell surrounds the insert's firebox, and room air flows between the insert body and the outer shell, where it is heated before being returned to the room. Some models have built-in fans to circulate the air. Most of the heat is delivered to the room rather than being trapped behind the insert in the fireplace brickwork. A decorative faceplate covers the space between the insert body and the fireplace opening.

A hearth-mount stove is another option for upgrading the performance of an older masonry fireplace. A hearth mount is a woodstove mounted in front of or partly inside the fireplace and vented through the fireplace throat. It must also be vented through a stainless-steel liner to the top of the chimney. Only certain woodstoves can be used as hearth mounts; check the certification label and installation instructions to see whether the model you have in mind can be vented through a fireplace.

When inserts and hearth mounts first came on the market in the early 1980s, they had a bad reputation for being unsafe, inefficient and expensive to maintain. Most inserts were not connected directly to the fireplace chimney, which allowed exhaust gases to leak from the flue collar and find their way up the chimney. They were fussy to light, smoky to use and very costly to maintain, since the insert had to be removed in order for the chimney to be cleaned.

In 1991, the Canadian code for installing wood-burning appliances (see "Installing a Wood-Burning Appliance" on p. 93) stipulated that every fireplace insert must have a corrosion-resistant, stainless-steel liner installed from the insert's flue collar all the way to the top of the chimney, providing a straight path up through the fireplace throat for a reliable draft and safer firing. (In the United States, no such national regulations exist. However, most municipal building codes require chimney liners to be installed with inserts.) Cleaning the liner is a simple matter of removing the chimney cap and running a brush down to the insert. Any deposits are cleaned from the firebox by removing the baffle. In the 20 years since this change in the Canadian code, fireplace inserts have become one of the most trouble-free of all wood-heating installations. Correctly installed, an advanced-technology fireplace insert can be almost as efficient as a freestanding wood-burning stove.

Fig. 3.6. Fireplace insert

Pros

- A fireplace insert converts an inefficient conventional fireplace into a highly efficient heater without sacrificing any of the charm of burning firewood.
- Aesthetically pleasing, a fireplace insert is handsome enough to be the centerpiece of a room.
- Cleaning an insert liner is easier than cleaning a chimney.

Cons

- Due to code requirements, there are high installation costs for stainless-steel liners, which may run as much as the insert or hearth-mount stove itself.
- Installation is complicated and is best left to a professional contractor.

Masonry Heater

Originating in ancient Rome and boasting a long history in Northern Europe after its widespread adoption there in the 1700s, the masonry heater operates on a different principle than the other advanced-technology wood-burning appliances: It uses tons of mass in the form of bricks or stone to store up and later release the heat it produces.

Instead of round-the-clock stoking and long firing cycles, the masonry heater relies on very short, very hot fires—above 2,000 degrees F (1,100°C). Its clean-burning fire charges the masonry mass, which radiates the stored heat for up to 24 hours, depending on weather conditions.

Also known as a "Russian heater," or *kachelöfen* (German for "clay oven"), the masonry heater has a core built from high-temperature firebrick that forms the firebox and heat-transfer passages. Brick or stone surrounds the core. A series of channels built into the masonry mass stores up to 80 percent of the hot gases produced by the fire. Compared with a conventional fireplace, which may send half of the heat produced up the chimney, a masonry heater is arguably the most effective and efficient wood heater money can buy. Because it burns so cleanly, chimney cleaning is a snap.

The downside to the masonry heater is its size and cost. It is extremely heavy, with an average footprint of 3 by 5 feet (1x1.5 m), and the cost of installing a masonry heater in an existing home is prohibitively expensive. New homes, however, can easily be built to accommodate one. A new masonry heater costs between US$5,000 and US$10,000 before installation. To ensure a safe and successful installation, the services of a mason certified by the Masonry Heater Association of North America are recommended. The additional cost will be amortized over many years, as the homeowner avoids the cost of frequent chimney cleanings and replacement catalytic converters and spends much less annually on firewood.

Pros

- A masonry heater is highly efficient, consuming only 25 to 50 percent of the firewood used by a wood-burning stove or fireplace.
- It produces a gentle, even heat.

Fig. 3.7. Masonry heater

- It is aesthetically pleasing to the eye.
- Less firewood burned means less chimney cleaning.

Cons

- A masonry heater is expensive, costing from US$5,000 to US$10,000, not including installation.
- It is too large to be practically installed in existing homes.

Central-Heating Systems

A central-heating system uses a network of air ducts or water pipes to distribute heat to all the rooms in your house. A furnace heats air and forces it through ducts with a fan. A boiler heats water and pumps it through pipes to heat radiators or floors.

Central heating systems that use a wood-burning furnace or boiler are not as common as they used to be. Older wood-burning furnaces or boilers were not as sophisticated as today's technologically advanced models, and they leaked a lot of heat. When homes became more energy-efficient and easier to heat with stoves and fireplaces that also provided a view of the fire, wood-fired central-heating systems fell out of fashion.

That said, the increased popularity of in-floor radiant heating via a network of pipes installed below the floor surface has led to a rekindling of interest in central heating with wood. A big advantage of these systems is that they can readily be used to heat domestic wash water in addition to providing heat for the house. Nevertheless, they have their detractors, as well as some pretty big disadvantages, particularly when they are located outdoors.

Wood-Burning Furnace

A large, sprawling house that cannot be effectively heated with a woodstove is usually a good candidate for a wood-burning furnace. This furnace typically does not require a lot of attention. While it must be stoked, a modern unit can do everything except ash removal with a simple setting of the thermostat. It is ideal for those who want the cost savings of heating with wood but not the mess and hassle of maintaining a wood-burning stove or a fireplace on the ground floor.

A wood-burning furnace forces either hot air throughout the ductwork or hot water through the pipes and radiators of your house. The advantages to this system are that the mess of tending a fire stays in the basement and that when properly installed, the furnace can heat your entire house. This is the best home-heating system for houses with many distinct rooms and a dearth of open space. A wood-burning furnace can also be installed as an add-on unit to work in conjunction with systems that use other fuels, such as oil, natural gas and electricity. Combination wood/oil or wood/electric furnaces using both energy sources in a single unit are also available.

The biggest drawback of wood-fired central heating is the price. It's the costliest of all wood-burning installations, as the services of a professional are absolutely necessary—particularly if a pairing with natural gas is involved. It may even require the installation of a separate chimney if the existing one is not approved to handle the new furnace. A wood-burning furnace is also dependent on electricity and is not as technologically advanced as are newer woodstoves and fireplace inserts, and its diffuse heat is no match for the cozy warmth of a fireplace.

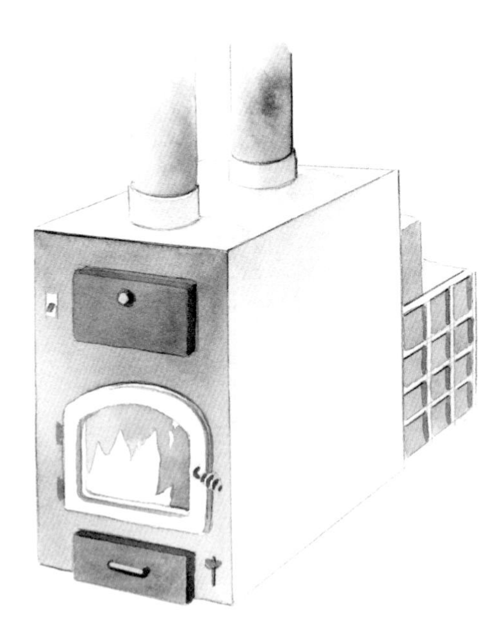

Fig. 3.8. Wood-burning furnace

Pros

- A wood-burning furnace is a good fit for older, larger, less fuel-efficient homes or homes with many small rooms and no large, open spaces.
- It restricts the mess of firewood and cleaning ashes to the basement and away from the living area.
- It can be combined with other home-heating systems, such as oil, coal and natural gas.

Cons

- A wood-burning furnace is expensive, requires a professional installation and may need a new chimney.
- It is dependent on electricity.
- It is not as cozy as a wood-burning stove or fireplace.

Outdoor Boiler

The concept of an outdoor wood-burning boiler certainly has its appeal. Locate the untidy business of fire husbandry and the thrumming noise of the furnace in a small insulated shed away from the house and run water pipes underground to transfer heat for both space and water heating. The past 20 years have seen a steady increase in the number of people in rural and semirural areas installing outdoor boilers. Wood-burning boilers have one major downside: Most models tend to produce prodigious amounts of dense smoke, so much so that some small towns have enacted bylaws restricting their installation by either banning them from residential areas or placing limits on their proximity to property lines.

The main reason the outdoor boiler is such a notorious smoker is that the firebox of most units is fully surrounded by a water jacket. While this is great for heating water, it makes smoke-free burning of the wood just about impossible. Complete combustion cannot occur below about 1,000 degrees F (540°C). Water at 150 degrees F (65°C) circulating around the boiler will quench the flames well before combustion is ever complete. Another reason for the smoke is the boiler's cyclical operating pattern. When the water temperature falls below a set point, the boiler's combustion air damper opens and a small fan forces combustion air into the firebox. Once the water is reheated to the set point, the fan shuts off and the damper closes. During these "off" cycles, the fire smolders and a huge cloud of smoke pours from the stack for about 10 minutes before the system settles back into a "normal" smoky fire. Mismatching the boiler size to your house only makes matters worse. An outdoor boiler connected to a modest-sized house will spend most of its time in the off cycle, particularly in fairly mild weather, so when it does fire, it is likely to create big plumes of smoke.

Older models are also woefully inefficient. The Canadian government has tested the overall efficiency of several types of outdoor boilers and has found them to be less than 50 percent efficient. In the United States, testing data obtained from manufacturers in 2008 by the Attorney General of New York State show that most outdoor boilers have heating efficiencies ranging from 28 to 55 percent, with an average of 43 percent. Some manufacturers have made an effort to offer their customers a cleaner and more efficient combustion system and are proving it by having their outdoor boilers emissions tested and certified. If you're thinking of buying an outdoor boiler, consider only those that are certified to meet EPA Phase 2 or CSA B366.1-1 standards for wood-burning appliances.

Fig. 3.9. Outdoor boiler

Pros

- An outdoor boiler keeps firewood and ashes outside and away from the living area.
- It can provide domestic hot water as well as water for home heating.
- It can provide hot water to several buildings, including sheds and garages.

Cons

- Older models are highly inefficient and produce massive amounts of nuisance smoke.
- Not many newer models are EPA-/CSA-certified.

Burn It Smart, Burn It Right

With the motto "Learn before you burn—burn the right wood, the right way, in the right appliance," the U.S. Environmental Protection Agency's Burn Wise offers partnership programs for wood-burning appliance manufacturers as well as public-awareness materials for consumers, air-quality agencies, chimney sweeps, hearth retailers and others who want to promote safe, healthy and responsible burning of wood for home heating.

The Burn Wise website (epa.gov/burnwise/) provides information on how to buy, install and maintain wood-burning appliances and fireplaces and suggests ways to improve community air quality through change-out programs and education. There are also details about how manufacturers can partner with the EPA to bring cleaner-burning appliances to market. Visitors can download case studies, recorded webinars and public-awareness materials, such as posters, brochures, tip sheets and website widgets. There are resources for tribal communities as well and an interactive map listing wood-burning ordinances and regulations by state.

Burn It Smart was a Canadian public-information program sponsored by the federal government from 2002 to 2010 and was facilitated by industry groups and/or other levels of government. Its objective was to help individuals burn wood safely and efficiently while keeping their families and communities healthy by reducing wood-smoke pollution. A series of public workshops were held in many Canadian towns and cities as part of the campaign. Today, provincial and private-sector agencies sponsor occasional workshops that deliver the original content under the Burn It Smart logo.

A typical workshop introduced participants to the carbon cycle and the concept of renewable energy resources. The workshop then delved into the efficiency differences between conventional wood-burning stoves (i.e., "old" potbellied, cast-iron box and cooking stoves) and fireplaces and the newer, advanced-technology CSA- and EPA-certified stoves and fireplace inserts. There was shoptalk about what to look for in a wood-burning appliance, the differences between indoor and outdoor chimneys, residential codes and safe installation techniques.

But the heart of the workshop was people sharing experiences and best practices and learning from one another about everything from burning dos and don'ts and how to keep indoor smoke to a minimum to controlling heat output and selecting, buying and storing firewood.

While the Burn It Smart program has been officially put out, many associations continue to hold information-awareness sessions on wood burning. Check with your local wood-heat retailer for more information.

Chapter 4

The Chimney

IF YOU BURN WOOD, YOU NEED A CHIMNEY. No wood-burning stove, fireplace or furnace can function properly without one. Indeed, most complaints about poorly performing wood-burning stoves can be traced back to chimney performance. The same fire in the same stove burning cleanly and efficiently in one house can smoke and smolder in another, solely because of how the chimney behaves. Knowing how a chimney functions is crucial when determining which type of wood-burning appliance to install and when working with it day to day.

A chimney operates on the principle that hot air rises. Hot gas rises in a chimney because it is less dense than the cold air outside the house. The rising gas creates a difference in pressure that is called draft, which sucks combustion air into the wood-burning appliance and expels the gas outdoors. (Wood-burning stoves and fireplaces have a natural draft, so they don't need a fan to push the exhaust gas up the chimney.) The hotter the temperature of the rising gas compared with the temperature outside, the stronger the draft. Put another way, think of the chimney as the engine that drives your wood-heating system. Think of heat as its fuel and draft as the horsepower it generates. The more heat you feed the chimney, the stronger the draft it produces. In a wood-heating system, draft is something you can't do without—it provides the suction that keeps smoke from spilling into your living room.

For the chimney to function properly, it has to receive about 20 percent of the heat energy produced by the burning wood. If the heat energy is below this level, water vapor can begin to condense in the chimney, especially one that is made of masonry and built outside the building envelope. That is why insulation in the chimney is so important—it helps keep the exhaust gas hot until it is expelled outside, thus increasing the draft.

A well-designed and properly installed chimney creates draft and pulls air up, even when no fire is burning. When you build a fire in a stove that is connected to such a chimney, your kindling ignites easily, the draft increases quickly and you should have a bright, hot fire right away with little or no smoke. That is the kind of chimney you want in your house.

Choosing a Chimney Type

Whether you're installing or upgrading a wood-burning appliance or building a new home that will feature a wood-heating system, the chimney will be made of metal or masonry. These are the two basic chimney materials approved for use with wood-burning systems.

Metal chimneys. Most building codes now require that wood-burning stoves, central-heating furnaces and some factory-built fireplaces use the 1,200-degree-F (650°C) standard developed in 1981. These chimneys are tested to withstand a higher continuous flue-gas temperature than are chimneys intended for other fuels. Most factory-built fireplaces are also approved for use with a special chimney that has the same upgraded liner found in the 1,200-degree-F (650°C) chimney. Your wood-heat retailer can show you the differences between these types and help you choose the most appropriate one for your situation. It goes without saying that all factory-built metal chimneys must be installed according to the manufacturer's instructions.

Masonry chimneys. If built to code, these traditional chimneys can be used with wood-burning appliances. They have a clay tile liner surrounded by a brick or stone shell. If you are building a masonry chimney, you will require a building permit, and make sure the mason who builds it is certified and follows building code rules. In a new masonry chimney, an insulated stainless-steel liner can be used as an alternative to the clay liner to reduce heat loss and improve performance.

Unsuitable Chimneys

A Type A chimney is an older class of metal chimney that was popular in the 1960s. It was originally designed for use with oil furnaces. The Type A cannot withstand the high temperatures of a chimney fire. If your chimney has a painted exterior or if the outside casing is square, it could be a Type A. If so, you must upgrade it to the 1,200-degree-F (650°C) standard with a liner before installing a new fireplace or wood-burning appliance.

A bracket masonry chimney is made of brick and built on wooden supports within a wall of the house. Common in older, rural homes, this chimney is not supported by adequate concrete foundations and cannot be upgraded to meet current building-code requirements. It should not be used.

An unlined masonry chimney should never be used for a wood-burning system, as building codes require that all chimneys be lined with clay tile, thermal concrete or stainless steel. In some cases, an old unlined chimney can be upgraded by installing a certified stainless-steel liner, assuming the chimney itself is sound.

An air-cooled chimney is installed with some decorative factory-built fireplaces and uses the flow of air between its inner and outer layers to keep the outer surface cool. The heat output from a wood-burning heating appliance, such as a stove or furnace, is too high to permit its use with an air-cooled chimney. These appliances should never be connected to an air-cooled chimney.

Assessing an Existing Chimney for Wood-Burning Safety

If your house has a recently built chimney and you're thinking of installing a wood-burning appliance, advise your insurance company. Then hire a chimney sweep certified by Wood Energy Technology

Transfer (WETT) in Canada or the Chimney Safety Institute of America (CSIA) in the United States (see "Installation Safety" on p. 108) to evaluate the chimney and confirm that it's safe for the planned installation. If you have a masonry chimney and there is deterioration of the bricks or mortar joints near the top of the chimney or dark stains on the brickwork, have it inspected immediately. Here are some of the questions the chimney sweep should address:

- Is the liner sound? If it has a clay liner, are there cracks, loose joints or deterioration?
- If the chimney is factory-built, is it sound, with proper clearances?
- Is the chimney suitable for the intended wood-burning appliance or system?
- Is the chimney flue going to be used for a single appliance only?
- If there is a pass-through (where the chimney and wood-burning appliance are on opposite sides of a combustible wall), are there adequate clearances?
- What is the condition of the fireplace? Is it up to code?

If your home was built before 1950, chances are, the chimney is not lined. On a bright day when the sun is high, shine a light up the chimney from either the cleanout or the fireplace. An unlined chimney will look like a brick wall; a lined chimney of this vintage will have a smooth surface of clay tiles, with all joints running horizontally. Don't assume that the entire chimney is lined or that it is sound enough to use without undergoing any repairs. Over time, chimney fires, water and the settling of the house can cause the mortar between the joints to fail.

If your chimney needs to be lined, building codes permit only three materials to be used: clay tile, thermal concrete and stainless steel.

Clay tile is relatively inexpensive to purchase but very difficult to install. If it isn't installed properly, it can lead to dangerous cracks into which creosote can work its way (see "Creosote: The Black Plague of Chimneys" on p. 88), which puts your chimney at risk of a chimney fire. Installing clay tile also requires the services of a qualified mason.

Thermal concrete is a ready-mix solution whereby an inflatable bladder is lowered into the chimney and a special insulating concrete mixture is poured around it. Once the concrete is set, the bladder is

Six Simple Rules for a Safe Chimney

Once installed or upgraded, your chimney requires regular maintenance and careful use to remain safe and prevent creosote buildup and indoor seepage of dangerous carbon monoxide. These six simple rules go a long way to keeping your chimney running cleanly and protecting your investment:

1. To each wood-burning appliance, a chimney. There are risks in venting more than one fireplace or wood-burning appliance to a single flue, ranging from the inconvenient (inadequate draft) to the dangerous (exhaust gases and carbon monoxide leaking back into your house). Remember, all vents to the outdoors, whether from other wood-burning appliances or heating systems, are technically chimneys and should be treated as such.

2. Schedule an annual chimney checkup. Every spring or in the fall, before the heating season, arrange an inspection and cleaning of your chimney with a chimney sweep who is certified by Wood Energy Technology Transfer (WETT) or the Chimney Safety Institute of America (CSIA). This will keep you on top of your chimney's cleanliness and performance.

3. Burn seasoned hardwoods in small, hot fires. Hardwoods burn longer, produce more heat and leave the least creosote behind. Use softwoods sparingly or only in spring or fall, when the demand for heat is less. Small, hot fires burn the smoke off before it gets to the chimney, leaving less creosote. Don't let it smolder!

4. Never starve your chimney of heat. Your chimney needs heat to function properly. Don't be afraid to burn the stove or fireplace hot, at least for a brief period, each time you start a fire or rekindle one using coals. This practice heats up both the appliance and the chimney structure and coaxes a dependable draft from the chimney for the rest of the heating cycle.

5. Evict pesky tenants. Watch what goes up your chimney, but look out for what comes down too. In addition to water, which can seep into a chimney and do structural damage to the liner or bricks and mortar, animal nests and other debris can block the flue and cause harmful gases to back up into your living area. Cap your chimney to keep the critters out.

6. Beware of mild winters. As pleasant as they may be, high-temperature winters mean longer, cooler fires with more smoke, which can swiftly increase the amount of creosote deposited in your chimney. If the last winter was mild and the stove didn't run often, you might think you can skip having it inspected or cleaned this year. You would be wrong.

removed, leaving the chimney with a smooth, seamless liner. A thermal concrete liner keeps the flue warm and clean but is expensive to install, with the cost running between US$2,000 and US$4,000.

Stainless-steel chimney liners come in both rigid and flexible forms. The rigid liner is made of sections of flue pipe held together with pop rivets and is installed with special tools; the flexible liner is similar to the aluminum ducting used to vent your household dryer. Either liner is placed in the center of the chimney, and an insulating mix of concrete is poured around it to hold it in place. Homeowners can install either type relatively easily, and both are comparatively easy to clean. For a straight chimney run, a rigid stainless-steel liner is best. The flexible stainless-steel liner is more expensive but can be snaked around bends and jogs in a chimney, which can offset the potential cost of having to break into your chimney to install the liner.

Planning a New Chimney Installation

If you're building a new house and plan to install a wood-burning system, save yourself a lot of frustration, discarded plans and contractor bills by planning the house around the system rather than the other way around. A chimney is a long-term investment, so it pays to get the right kind, follow a carefully considered plan and have the installation done according to code. Most wood-heat retailers and chimney sweeps can offer guidance; there are also government agencies and publications to which you can refer. In Canada, the best strategy is to have your chimney installed by a professional certified by Wood Energy Technology Transfer (WETT). In the United States, look for a tradesperson who has been certified by the Chimney Safety Institute of America (CSIA).

Here are several key considerations to keep in mind when preparing for the installation of a new chimney:

Clear the house. Building codes require that the top of the chimney extend at least 3 feet (1 m) above the point where it exits the roof and 2 feet (60 cm) above any roof surface or structure within a horizontal distance of 10 feet (3 m). These rules are intended to position the top of the chimney above any areas of air turbulence caused by wind (see Fig. 4.3 on p. 86).

Pushing the Envelope: How Not to Locate a Chimney in a House

Why do outside chimneys allow cold air and odors to leak into the house when there is no fire burning in the stove or fireplace? When it's cold outside, warm air rising in the house creates an area of low pressure in the lower part of the house and an area of slightly higher pressure in the upper part. A neutral pressure plane divides these zones. If there is no fire burning inside and the air in the chimney outside cools to below room temperature, the chimney actually reverses its draft and sucks cold, smelly air down the flue (see Fig.4.1). This is called the "stack effect."

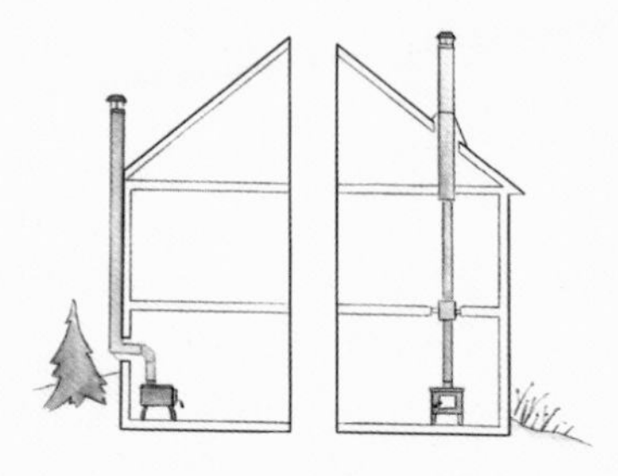

Fig. 4.1. *A chimney installed outside the house (left) is vulnerable to the stack effect in cold temperatures. A chimney installed inside the building envelope (right) provides a constant standby draft.*

A chimney installed inside the house envelope is not affected by the stack effect because it is always at or above the indoor temperature. A neutral pressure plane always moves toward the biggest leaks, so the opening at the top of the chimney has a higher neutral pressure plane than the opening in the stove or fireplace in the lower part of the house. This creates a gentle upward movement of air through the chimney when there is no fire burning. This is known as "standby draft," and the resulting low-pressure zone at the fireplace opening means that when you light a fire, it will kindle quickly without smoking.

You can further improve your chimney's performance, as well as that of your fireplace or wood-burning stove, by installing your hearth away from outside walls and closer to the heart of your home. The chimney then rises through warm space for most of its length and exits near the roof's peak. The result is a well-drafted chimney, unaffected by cold temperatures or harsh winds (see Fig. 4.2). Don't make the mistake of locating a chimney low on the eaves of a cathedral roof or in a one-story section of a two-story house or in an area where the chimney exits the roof below the top of the house. In any of these configurations, as surely as heat rises, cold air will spill from the appliance when no fire is burning.

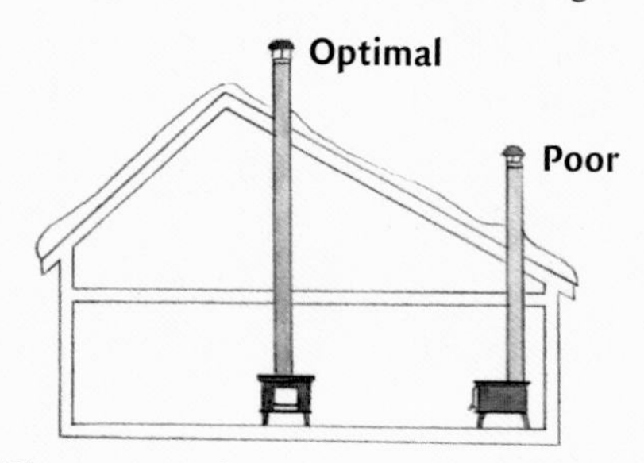

Fig. 4.2. *Stove location*

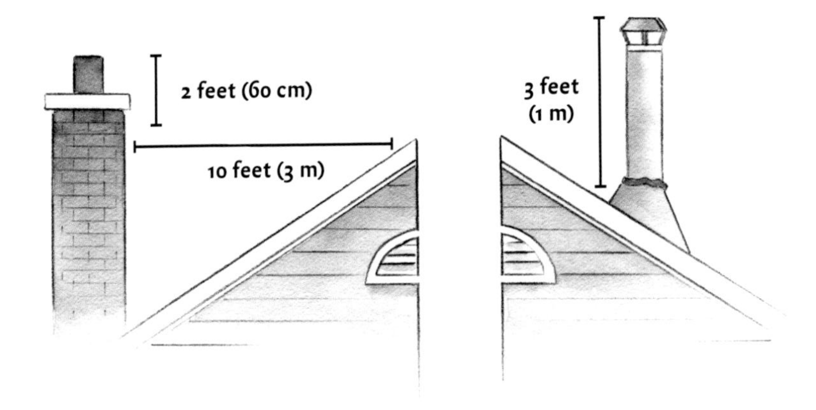

Fig. 4.3. The top of a chimney should be higher than any air turbulence caused by wind blowing against the roof.

Keep the chimney inside the house. The interior of the house is warm, and you want to locate the chimney where? The biggest blunder you can make when installing a new chimney is to locate it in a chase outside the building envelope to conserve interior space. When a chimney runs up an external wall, it is exposed to seasonally cold temperatures. That can create cold backdrafts when the appliance is not in use, making it hard to build a fire without spilling smoke into the house. A chimney that runs up through the house benefits from being enclosed within the warm house environment and produces a stronger, more reliable draft (see "Pushing the Envelope: How Not to Locate a New Chimney in a House" on p. 85). Early North American settlers knew a few things we have long forgotten. Take a hint from our forebears—locate your chimney inside the house, if possible.

Match the chimney to the appliance. A high-performance wood-burning stove or fireplace demands a high-performance chimney. If you want to install a purely decorative fireplace without gasketed doors and air controls, a chimney that isn't insulated (i.e., a conventional masonry chimney or an air-cooled chimney) is acceptable. But if you have an advanced-technology wood-burning stove or fireplace, you'll need a high-performance, 1,200-degree-F (650°C) model and a specialized fireplace chimney or lined masonry chimney.

Keep it straight. Each turn of a chimney creates resistance to exhaust gas flow through the flue and lowers the draft. A reliable rule of thumb? Each 90-degree turn cuts 5 feet (1.5 m) from the effective height of a chimney.

Go tall or go home. *Ceteris paribus*—all other things being equal—a tall chimney has a stronger draft than does a short one. The total system height (from the floor on which the appliance is mounted to the top of the chimney) should never be less than 15 feet (4.5 m). A fireplace with a large opening or with more than one open side, such as see-through or corner fireplaces, benefits from a taller system height. Don't cut corners on height—it's easier to decrease the draft in a tall chimney than to boost the draft in a short one.

Banish the corners. A round chimney flue drafts better than a square or rectangular one, as the corners offer a place for exhaust gases to pool and condense. In addition, an insulated chimney (lined with either metal or masonry) grows warm faster and stays warm longer, making it easier for the chimney to draft properly.

Follow the rules. Be sure to satisfy all clearances and safety and building codes. These are required for good reason, and not meeting or exceeding the standard can be very costly should your new wood-burning system fail or, worse, cause a fire in your house.

Keep your hearth central. A chimney should penetrate the highest part of the building envelope for maximum draft. After making sure to locate the chimney inside the house, this is perhaps the key to operating an efficient wood-burning stove or fireplace. For most common house designs, this means that the fireplace and chimney should be located on an interior wall rather than an outside wall (see "Pushing the Envelope: How Not to Locate a New Chimney in a House" on p. 85).

Creosote: The Black Plague of Chimneys

Creosote is condensed wood smoke, a black, tarry, foul-smelling, corrosive and highly combustible substance that, if left unchecked, coats the insides of everything it passes through. It forms when tar droplets, vapors and other organic compounds given off during the wood-burning process condense and solidify on their way out of the chimney as the smoke cools below 250 degrees F (120°C). While creosote has its uses—in meat preservation and treating wood and, medicinally, as an expectorant, an astringent and a laxative—it has no place in a chimney, where it is a nuisance that can have deadly consequences if proper precautions are not taken.

Creosote is always present in wood burning, but it takes different forms depending on the location of your chimney (interior versus exterior), the type of wood you burn (seasoned versus green) and the kind of fires you build (high heat versus smoldering). Hot, consistent fires using seasoned wood and a chimney located inside the house allow very little creosote to build up, and what does occur is flaky and powdery. Low-heat fires using wet, unseasoned wood and an exterior chimney create the perfect conditions for the formation of hard, glazed creosote.

People began to wake up to the dangers of creosote in the 1970s, when new airtight stoves offered the option of overnight fires that did not need to be fed every hour. But the air-starved, slow burn of these fires kept the chimney stacks cool, and creosote gradually lined the chimneys. Just as cholesterol contributes to hardening of the arteries, creosote deposits in a chimney can become several inches thick over the course of a heating season, compounding the problem. The deposits reduce the draft, which starves the fire of the oxygen needed to burn the wood at high temperatures to boost the draft—a vicious cycle.

Creosote is highly combustible. When a scorching fire is built in the stove or fireplace and the air control is left wide open, hot oxygen can escape into the chimney and ignite the creosote, causing a chimney fire. Every year, millions of dollars in damage occur in Canada and the United States as the result of chimney fires. At the very least, a chimney fire can irreparably damage the chimney liner; at worst, it can spread to the rest of the house, destroying property—and lives.

Cleaning Out the Creosote—When and How Often?

Your chimney and flue-pipe assemblies must be cleaned of creosote deposits on a regular basis to keep your fireplace, woodstove or furnace burning cleanly and safely. When and how often depend on the chimney and appliance type, the local climate and the frequency and type of fires burned, but a good guideline is to

clean whenever the deposits are greater than a quarter of an inch (6 mm) thick or every fall, whichever comes first. Never assume your chimney is clean. Check it regularly to be certain, especially during the spring and fall. If you do have a chimney fire, have the chimney inspected and repaired, if necessary, before using the system again. Forget about burning one of those magic cleaning logs. Hire a licensed chimney sweep certified by Wood Energy Technology Transfer (WETT) or the Chimney Safety Institute of America (CSIA) to inspect your chimney.

Can Creosote Buildup Be Prevented?
No. But by burning only seasoned wood (which produces less of the steam that allows wood smoke to condense in the chimney) in small, hot fires (which keeps the stack temperature above creosote's condensation point of 250 degrees F/120°C) in an advanced-technology wood-burning appliance (which burns the smoke before it even exits the fireplace), your creosote buildup will be of the benign flaky kind and shouldn't require more than a yearly cleaning.

To Cap or Not to Cap?

If you have an existing metal chimney or plan to install one as part of your wood-burning heating system, you should cap it for a number of reasons, the most important of which is to keep the rain out. Without a cap, rain will run straight into your wood-burning stove, fireplace or heater. A cap also prevents birds and other animals from building nests in your nice warm chimney. Because a metal chimney is made of several layers of metal, with air or insulation between them, leaving it uncapped allows water and moisture to seep between the layers, corroding the metal and damaging the chimney's structural integrity.

A masonry chimney, on the other hand, does not necessarily need a cap, but there are advantages to installing one. Caps for masonry chimneys are made from a number of materials, including stainless steel, copper and aluminum. Most caps feature side screening that prevents birds and small animals from getting into the chimney and helps keep in any large stray sparks.

Advantages

- Keeps animals, especially raccoons and birds, from setting up house in your chimney.
- Prevents rain and moisture from building up inside the chimney, which helps reduce the "chimney smell" that often invades homes when temperatures are mild.
- Sheds ice from the chimney crown, extending the life of the chimney structure and eliminating some causes of cracking in the flue tiles.
- Increases draft, especially in areas with constant wind.

Disadvantages

- An improperly designed or installed chimney cap can potentially slow a chimney's draft.
- Some chimney caps have mesh screening that can become clogged with soot and creosote after heavy use, causing draft reductions and poor stove or fireplace performance.
- In some cases, creosote can collect on the chimney cap and run down the exterior of the chimney, staining the stucco or brick.

This is usually a problem only with older woodstoves and/or when using unseasoned wood.

- If you live in an area prone to high winds, a poorly installed chimney cap can easily blow off, requiring a trip to the roof to reinstall it or a trip to the hardware store to replace it.

Chapter 5

Installing a Wood-Burning Appliance

AS NOTED BEFORE, CAREFUL THOUGHT AND preparation are critical if you are planning to heat your home with wood. Think long and hard about your level of commitment to the wood-burning lifestyle (see "Choosing a Wood-Heating Strategy" on p. 53), crunch the numbers, and do some homework about where to locate a wood-burning appliance in your home before buying one and having it installed. Whether you choose a woodstove, a high-efficiency factory-built fireplace, a masonry heater, a fireplace insert or a pellet stove, the same general issues pertain.

Put it where you live. Choosing the right location for your wood-burning appliance is one of the most important decisions you'll make. If the appliance is meant to provide the lion's share of your home heat, situate it where you and your family spend the most time. That is usually on the main floor, where the kitchen and living and dining rooms are located. By installing the appliance in this area, you will be warm and comfortable while preparing and eating meals and relaxing in the evenings.

Don't banish it to the basement. Unless you are installing a central-heating system, such as a wood-burning furnace, a basement location is a poor choice for effective wood heating. It's true that heat rises, but to heat the upper levels of your home efficiently, you will likely overheat your basement, burn through your woodpile at a prodigious rate and damage your stove through constant high firing. An unfinished basement is particularly unsuitable for a wood-burning appliance because the walls and floor absorb much of the heat generated. A wood-burning appliance is appropriate for a basement space only if it is located in a room where you spend a lot of time—say, a family or recreation room—and if the chimney is installed up through the house interior rather than up an external wall.

Match appliance size to room size. The layout of your house should be the deciding factor in your choice of wood-burning appliance. If the main floor is divided into small rooms, a wood-burning stove located in one room will overheat that room and fail to heat the others. Houses with small rooms are more effectively heated by a central-heating furnace or an outdoor boiler. If the main floor is an open-concept design, however, it can often be heated entirely with a fireplace or wood-burning stove. Because heat flows unimpeded to most of this area, a larger space heater can be used without causing overheating. Bring along blueprints or a floor plan of your house, and ask an experienced wood-heat retailer to recommend an appropriate model.

Consider the chimney. If you are installing a fireplace or masonry heater in a new home, chimney type, location and arrangement are integral to how effectively your wood-heating system will function. When choosing the location for the appliance, consider the placement and type of chimney. If at all possible, avoid running the chimney up an exterior wall of the house.

Move the heat around. For a wood-burning stove or fireplace to heat your home comfortably and efficiently, heat must flow from one area of the house to another. How easily this is achieved depends on the layout of your house and how it was built. An energy-efficient house with an open-concept design is easier to heat with a single stove or fireplace because the house loses heat slowly, even in bedrooms that are located a distance away from the appliance. A leaky house is the opposite: The rooms farthest from the heat source cool quickly. The good news is that in many cases, the heat circulated by natural convection can be distributed around the home. Any heat produced by your wood-burning appliance rises to ceiling level, moves across to the room's walls and sinks down the walls to the floor, where it is drawn back to the appliance to be reheated. Some heat collected at ceiling level seeps out of the room and rises to the upper levels of the house. If your stove or fireplace operates more or less continuously, you can direct these convection currents to improve the distribution of heat using the following techniques:

- If the woodstove is located in a room with a high ceiling, install a ceiling fan to push the heated air back down into the living space and throughout the house. Set the fan to turn counterclockwise (which looks like clockwise when you are standing directly under it) to redirect the warm air from the ceiling down the walls. The opposite setting is used to cool a room in summer.
- The wall area above interior doors impedes the flow of ceiling-level warm air into hallways and nearby rooms. If you don't mind the aesthetic result, install a small fan in the top corner of a doorway to redirect heat out of the room (these fans are available from wood-heat dealers).
- Use the air-circulating fan of a central-heating furnace to move air around the house. With the furnace fan set to low, the air heated by your wood-burning appliance is gradually distributed throughout the house. (The disadvantage of this method is increased electricity consumption, particularly if your furnace does not have a high-efficiency fan.)
- Install large grilles in walls and floors to create a path for the flow of warm air. The downside to grilles is an increase in noise between rooms and reduced privacy.

Installing a Wood-Burning Stove

Woodstoves are, by far, the most common type of wood-burning appliance in North America, accounting for more than half of all wood-heating installations. In addition, many homeowners choose to install and maintain their own stoves. For these reasons, this book offers more details on how to install a woodstove than on how to install other types of wood-burning appliances. That said, the following general guidelines are not intended to replace the advice and expertise of a wood-heat professional. If you intend to do your own installation, visit several appliance dealers and speak with experienced contractors. Look for individuals trained under the WETT program in Canada or the National Fireplace Institute (NFI) in the United States (see "Installation Safety" on p. 108).

Minimum Clearances: Keeping a Safe Distance

The most important consideration when installing any wood-burning appliance is establishing clearance between the appliance and any combustible materials. More than 50 percent of do-it-yourself woodstove installations are unsafe because the appliance is located too close to a combustible surface. The kindling temperature (the temperature at which walls, ceilings, floors, curtains and furniture become combustible) can drop over time due to exposure to the heat produced by the appliance, increasing the chance of ignition. These surfaces must be protected from any heat emitted by a wood-burning system. The simplest form of protection is to maintain the stove's safe minimum distance from combustibles or to install a shield that prevents the heat from reaching them.

Installation guidelines for woodstoves can be grouped into two categories: stoves that are CSA- or EPA-certified (see "Certified Wood-Burning Appliances" on p. 56) and stoves that are not certified. Current regulations in both the United States and Canada stipulate that all new woodstoves must be safety-certified. Minimum installation clearances and other guidelines for these appliances can be found in the manufacturer's installation instructions.

Stoves built before 1980 fall into the second category. You may have found or inherited a beautiful antique cooking range, but you should think twice before using it to heat your home. In addition to being less efficient than an advanced-technology wood-burning stove, an uncertified stove is far more complicated to install. According to Canadian

Standards Association (CSA) Standard B365-10 (Installation Code for Solid-Fuel-Burning Appliances and Equipment), the minimum clearance for an uncertified radiant stove is 48 inches (120 cm) to the sides and rear or 36 inches (90 cm) if the stove is surrounded by a sheet-metal jacket or casing behind which convection air can flow. In the United States, National Fire Protection Association (NFPA) Standard 211 (Standard for Chimneys, Fireplaces, Vents, and Solid Fuel-Burning Appliances) stipulates that this distance be no less than 36 inches (90 cm) on all sides (see "Installation Safety" on p. 108). (These clearances are overly large because they are intended to cover all shapes, sizes and designs of untested stoves and to provide a safe guideline for furniture, curtains, rugs and other living-area objects.) If the fact that by law, your stove has to be 4 feet (1.2 m) from the wall does not convince you of the importance of generous clearances, keep in mind that many insurance companies are reluctant to provide coverage to homes where uncertified stoves have been installed. If you still want to install that uncertified range, seek the advice of a qualified wood-heat dealer, an installer or a chimney sweep. Or, better still, consult all three.

Safely Reducing Minimum Clearances

Even with the small clearances required by CSA- or EPA-certified stoves, most homeowners want their wood-burning stoves to take up as little floor space as possible. Clearances to walls and ceilings can be reduced even more by using special permanently mounted heat shields. These can be made from simple sheet metal, but decorative brick, stone and ceramic-tile shields are also available. Canadian and U.S. clearances for certified and uncertified stoves are set out in Tables 5.1 and 5.2 (see p. 98 and p. 100). Simply refer to your appliance's label or installation instructions to determine the minimum clearance, then calculate the allowed clearance reduction for the type of shield you plan to use.

What these types of shields have in common is the air space they create between the shield and the combustible surface. When the stove is operating, this space sets up a convection flow of air and prevents the stove's heat from reaching the surface behind it. You need

Table 5.1

Reducing Clearances to Combustible Materials with Heat Shields (Canada)		
	CLEARANCES MAY BE REDUCED BY THESE PERCENTAGES	
TYPE OF HEAT SHIELD	SIDES AND REAR	TOP
Sheet metal, minimum 29 gauge, spaced out at least ⅞ inch (21 mm) by non-combustible spacers	67%	50%
Ceramic or equivalent non-combustible tiles on non-combustible supports, spaced out at least ⅞ inch (21 mm) by non-combustible spacers	50%	33%
Ceramic or equivalent non-combustible tiles on non-combustible supports with a minimum 29-gauge sheet-metal backing, spaced out at least ⅞ inch (21 mm) by non-combustible spacers	67%	50%
Brick, spaced out at least ⅞ inch (21 mm) by non-combustible spacers	50%	N/A
Brick with a minimum 29-gauge sheet-metal backing, spaced out at least ⅞ inch (21 mm) by non-combustible spacers	67%	N/A

Source: CSA Standard B365-10, Table 3: "Reduction in Appliance and Ductwork Clearance from Combustible Material with Specified Forms of Protection"

good channel spacers to provide solid support to the shield; metal wall strapping, available at most building-supply stores, works well. The shield must extend 18 inches (45 cm) beyond each edge of the stove and 20 inches (50 cm) above the top of the stove (see Fig. 5.1). Note that even if the stove clearance is reduced by using a suitable shield, flue-pipe clearances must still comply with the clearance rules.

Some shields allow you to reduce minimum clearances. These shields are put through a rigorous series of tests to determine how effectively they reduce clearances. They are certified and carry a label that provides clearance-reduction details. Some stoves come with built-in heat shields; always install these according to the manufacturer's instructions.

Fig. 5.1. A heat shield allows air to circulate between the shield and any combustible surface and permits the reduction of clearances for a woodstove.

Protecting the Floor

While CSA- or EPA-certified stoves are tested to ensure that they will not overheat a floor, you still need to protect the floor from any live embers that might fall from the stove during fire tending or ash removal. A floor pad (sometimes called a "stove board") is essential. The floor pad must have a continuous surface and be made of a durable, non-combustible material, such as sheet metal, ceramic tile or

Table 5.2

Reducing Clearances to Combustible Materials with Heat Shields (United States)		
	MAXIMUM ALLOWABLE REDUCTION IN CLEARANCE	
TYPE OF HEAT SHIELD	AS WALL PROTECTOR	AS CEILING PROTECTOR
3½-inch-thick (90 mm) masonry wall without ventilated air space	33%	—
½-inch-thick (13 mm) non-combustible insulation board over 1-inch (2.5 cm) glass fiber or mineral wool batts without ventilated air space	50%	33%
24-gauge sheet metal over 1-inch (2.5 cm) glass fiber or mineral wool batts reinforced with wire or equivalent, on rear face with ventilated air space	66%	50%
3½-inch-thick (90 mm) masonry wall with ventilated air space	66%	—
24-gauge sheet metal with ventilated air space	66%	50%
½-inch-thick (13 mm) non-combustible insulation board with ventilated air space	66%	50%

Source: NFPA Standard 211, Table 16.6.2.1: "Reduction of Appliance Clearance with Specified Forms of Protection"

brick. It must extend at least 18 inches (45 cm) in front of the loading door and 8 inches (20 cm) beyond the sides and back of the stove. Never install a floor pad on a carpet.

Uncertified stoves have never passed a safety test, and they may give off heat that could dangerously overheat floors. Stoves with legs shorter than 2 inches (5 cm) should not be installed on a combustible floor at all. The rules for floor protection for uncertified stoves are complicated, with several different types stipulated depending on the length of the stove legs. If you are installing an uncertified stove, contact a qualified professional for advice.

Flue Pipes: The Weak Link

Flue pipes, or stovepipes, carry the exhaust gases from your wood-burning stove's flue collar to the base of the chimney. These pipes are the weak link in a wood-burning system because they are all too often installed improperly. The two most common mistakes in flue-pipe installations are that they are installed too close to combustible surfaces or that they are not securely fastened and come apart when undergoing the rigors of daily fires. If you install a single-wall flue-pipe assembly for your new wood-burning stove, make sure it is installed exactly according to the list of safety rules provided here.

With flue-pipe assemblies, the straighter and shorter the path between the stove and the entrance to the chimney, the better. The ideal single-wall flue-pipe assembly rises straight up from the stove's

Fig. 5.2. A straight flue-pipe assembly offers the path of least resistance to gas flow and creates a stronger draft. Straight assemblies also require less maintenance, as there are no horizontal sections where creosote can build up.

flue collar and directly into the chimney through a metal thimble, with no elbows (smoke is like a car—corners tend to slow it down). If you must incorporate elbows into the assembly, try not to use more than two. Less is also more: The longer the length of pipe, the more creosote you may have to deal with as the pipe cools between fires. Most chimney sweeps recommend not exceeding 10 feet (3 m) of flue pipe in an installation.

If you must use a horizontal section of pipe, it should slope up toward the chimney at a rate of ¼ inch for every foot (2 cm/m). The pipe should never slope down. Install the pipe sections with the seams on top to prevent creosote from dripping out and creating a fire hazard. Similarly, install all vertical sections of pipe with crimped (male) ends pointing down to prevent leakage. Secure any joints with at least three sheet-metal screws for stability. And, unless you want to move the stove every time you need to inspect the pipes for creosote buildup, insert an inspection sleeve or a telescopic section into your assembly. As a bonus, this insert allows for some expansion, as the pipes grow hot during firing.

When buying stovepipe, note the gauge, or thickness, of the pipe. Lower is better, so 24-gauge pipe withstands daily use better than 26-gauge pipe and lasts longer, making up for its higher price. Never use galvanized flue pipe in a wood-burning installation. It gives off toxic zinc vapor at temperatures of 750 degrees F (400°C) or higher.

Rules for Single-Wall Flue-Pipe Assemblies

- Minimum clearance from combustible material/surface: 18 inches (45 cm). If suitable shielding is installed either on the pipe or on the combustible surface, the minimum clearance may be reduced by 50 percent.
- Maximum overall length of straight pipe: 10 feet (3 m).
- Maximum unsupported horizontal length of straight pipe: 10 feet (3 m).
- Maximum number of 90-degree elbows: two.
- Minimum upward slope toward the chimney: ¼ inch per foot (2 cm/m).
- Number of screws with which each joint in the assembly must be

fastened, including connections at the flue collar and chimney: three.

- Minimum gauge for 6-, 7- and 8-inch-diameter (15, 18 and 20 cm) flue pipes: 24.
- Allow for expansion of the assembly by including elbows, a telescopic section or an inspection sleeve.
- Crimped, or male, ends of all pipe sections must be oriented toward the appliance.

Double-Wall Flue-Pipe Assemblies

Certified double-wall flue pipes feature a stainless-steel inner liner and a sealed or vented outer shell. They are more expensive than single-wall pipes but last longer and produce a more stable assembly. Because the two walls offer an extra layer of insulation, they require smaller clearances than do single-wall pipes, and longer lengths of pipe can be used if stove placement is an issue. For the specific clearances allowed, check the labels attached to the pipes as well as the manufacturer's instructions.

Installing Other Wood-Burning Appliances

During the oil crisis of the 1970s, when many homeowners turned to firewood to combat skyrocketing home-heating bills, old pre-code fireplaces were often retrofitted with wood-burning stoves or fireplace inserts. Many of these installations did not achieve a tight seal between the stove's collar and the chimney's flue. Some homeowners simply shoved the stovepipe up past the fireplace damper and filled the remaining space with insulation. Others covered the fireplace opening with a metal plate and ran the flue pipe through a hole in it.

The problem with these homemade home-heating solutions was that smoke did not exit the chimney immediately but, rather, lingered in the fireplace, producing a large and dangerous buildup of creosote that often led to chimney fires. The only effective way to install a wood-burning appliance in an existing fireplace is to reline the chimney flue and connect the stove directly to the new liner. (Indeed, CSA code in Canada makes this mandatory. If you choose to install any wood-burning appliance in an existing hearth—whether it's a factory-built fireplace, a hearth-mount stove or a fireplace insert—a full stainless-steel chimney liner must be installed from the flue collar to the top of the chimney. In the United States, minimum requirement codes for installation do not exist at the national level, but most municipal building codes require that a chimney liner be installed with a hearth mount or an insert.) The liner matches the flue to the size of the insert collar and isolates the flue gas from the surrounding masonry structure, retaining heat and producing a stronger draft. It also makes cleaning and servicing easier, since the liner can be cleaned from the top of the chimney. With a full liner, the insert does not have to be removed to clean the liner, a procedure that is costly and potentially damaging to the hearth.

Installing Fireplace Inserts and Hearth-Mount Stoves

The installation of a fireplace insert or a hearth-mount stove and its chimney liner is permanent. The structure of the masonry fireplace must be altered to complete the installation, and it may not be possible to return it to its original condition. Also, in most cases, the hearth has to be extended at least 18 inches (45 cm) beyond the front of the appliance to provide protection for the floor, an extension that must be permanently mounted. (No matter what manufacturers claim, hearth rugs are simply not adequate.) While installing a fireplace insert may look relatively easy, it is far from a do-it-yourself job. For one thing, the existing fireplace and chimney must first be cleaned thoroughly so that no creosote deposits remain. In short, installing a fireplace insert or a hearth-mount stove is a job best left to a professional.

Installing Factory-Built Fireplaces

Also known as "zero-clearance" fireplaces, factory-built fireplaces are prefabricated metal units designed to be enclosed—back, top, bottom and sides—by the body of a newly built fireplace. Because these fireplaces are installed against and surrounded by combustible building materials, great care must be taken to install them according to the manufacturer's exact specifications. Factory-built fireplaces cannot safely accept fireplace inserts or after-market glass doors. If

you have such a fireplace, use it only according to the manufacturer's instructions, and do not attempt to retrofit it.

Installing Masonry Heaters

A masonry heater is entirely different from a conventional fireplace in design, construction and operation. It must be assembled on site by a certified mason—simple errors made by an inexperienced installer can result in the heater's failing to work once it is built in place. The clearances of a masonry heater to surrounding combustible construction must meet the requirements outlined in the local building codes for a conventional fireplace. The Masonry Heater Association of North America (MHA) sponsors a heater-mason training and

certification program. To guarantee that your masonry heater is effective and durable, be sure it's designed and built by an MHA-certified heater mason. (To find a certified heater mason in your area, consult your local wood-heat dealer.)

Installing Pellet Stoves

Installation guidelines for pellet stoves vary widely. The manufacturer's instruction manual is the best resource for details regarding clearances, appropriate materials to be used for exhaust venting and the arrangement of vent components. Because a pellet stove uses an internal fan to draw in combustion air and to force the exhaust into the venting system, each joint must be carefully sealed to prevent exhaust leakage into the house. Unlike a wood-burning stove, a pellet stove is not installed with flue pipe. Instead, it must be connected to a proper venting system, and the combustion fan and auger must be adjusted to burn the pellets correctly. Your best assurance of a successful installation is to have it done by an experienced professional.

Installing Central-Heating Furnaces and Boilers

This is easily the most complex installation of a wood-heating system for your house. If you are considering converting to wood heat using a central heater or a boiler or if you're building a new home with such a system in mind, your heating retailer or contractor is the best source of information. The installation of a central-heating appliance is complex and requires several specialized skills. Hire professionals to do the work. Installation rules for each model of central-heating furnace or boiler can vary, so consult the manufacturer's instruction manual.

Cleaning and Maintenance

Once your wood-burning appliance is safely up and running, it requires regular maintenance and cleaning to ensure that it continues to provide heat efficiently and effectively for years to come.

Adjusting doors. To make sure the seal on the loading door of your wood-burning stove does not leak, open it when the stove is cold,

Installation Safety

A safe wood-burning system features all of the following items:

- A CSA- or EPA-certified stove, fireplace or furnace.
- The right type of chimney for the appliance.
- A system design that adheres to code and avoids safety compromises.
- Reliable advice on safe installation or, preferably, installation by a qualified professional.

Prior to the 1980s, wood-burning appliances were not rigorously tested for safety and homeowners had little or no guidance in how to install them. As a result, many do-it-yourself installations led to deadly carbon monoxide poisoning and house fires. Since that time, there has been a concerted effort by all levels of government and the wood-heating industry in Canada and the United States to improve the safety record of residential wood burning. A number of measures have been implemented to help homeowners heat with wood safely.

- Reliable installation codes were developed in 1980. In Canada, see CSA Standard B365, Installation Code for Solid-Fuel-Burning Appliances and Equipment (the most recent version, B365-10, was published in 2010). In the United States, see NFPA 211, Standard for Chimneys, Fireplaces, Vents, and Solid Fuel-Burning Appliances (the most recent version was published in 2013).
- Product safety standards for stoves, inserts, fireplaces, furnaces, chimneys and flue pipes have been developed.
- Training programs have been created for wood-heat retailers, contractors, chimney sweeps, municipal fire and building inspectors and insurance inspectors—the WETT Wood Energy Technical Training program in Canada and National Fireplace Institute (NFI) certification in the United States.

Canada: WETT

Wood Energy Technology Transfer Inc. (WETT) is a non-profit training and education association dedicated to promoting the safe and effective use of wood-burning systems in Canada. Managed by a volunteer board of directors elected by holders of valid WETT certificates, the association offers the Wood Energy Technical Training program, a series of courses for contractors, chimney sweeps, wood-heat retailers, municipal fire and building inspectors and insurance inspectors to learn the details of safety codes and the proper installation, maintenance and inspection of wood-burning equipment and systems. Graduates of the program are issued certificates and are allowed to display the WETT logo in their stores, on their service vehicles and in advertising materials. WETT does not certify businesses but, rather, the individuals who pass its courses and have adequate field experience. Look for the WETT logo when seeking reliable wood-heat information, advice and installation or maintenance services.

United States: NFI

The National Fireplace Institute (NFI) is the professional certification division of the American Hearth, Patio & Barbecue Education Foundation. A non-profit educational organization for the hearth industry, the NFI was formed in 2001 with a mandate to increase public safety by establishing credentials for professionals involved in the planning and installation of residential hearth appliances and heating systems. The NFI certifies three types of specialists: Gas Specialists, Woodburning Specialists and Pellet Specialists. Individuals who have all three NFI certifications are Master Hearth Professionals. To become certified, individuals must take exams covering core knowledge such as combustion and heat transfer, as well as information about the relevant appliances and venting systems. Many wood-burning appliance manufacturers in the United States promote NFI certification under the NFI Advocacy Program.

Before you make any decisions about your hearth, visit at least two specialty wood-heat retailers and ask a lot of questions. Look for WETT or NFI certificates. Speak to your insurance agent, and ask others who

have installed wood-heating systems of their own. Only then will you be able to get a balanced view of how to match your needs and objectives with a wood-heating system that will achieve them.

Safety Checklist

Stove

- ❑ Adequate clearances between stove and all combustible materials
- ❑ Structural integrity: Look for cracks and warping; check functional moving parts and catalytic converter
- ❑ Tightness of door latches, seals and gaskets

Stovepipe

- ❑ Adequate clearances between stovepipe and all combustible materials
- ❑ Pipe configuration conducive to safe, clean burning; length not to exceed 8 feet (2.4 m)
- ❑ Sheet-metal screws at each joint of pipe; support brackets on horizontal sections of 4 feet (1.2 m) or more

Fireplace

- ❑ Adequate clearances between fireplace and all combustible materials
- ❑ Structural integrity of mortar, firebricks and damper
- ❑ Functioning fireplace screen and/or glass doors

Chimney

- ❑ Structural integrity of liners, walls, crown, mortared joints, thimbles and clean-out doors
- ❑ Cleanliness
- ❑ Separate flue for each appliance

Detection Devices

- ❑ Working smoke alarms on every floor of the house
- ❑ Working carbon monoxide detectors on every floor of the house
- ❑ Charged fire extinguisher in close proximity to each wood-burning appliance or fireplace

Wood

- ❑ Seasoned
- ❑ Stored safely clear of wood-burning appliances and fireplaces
- ❑ Safe storage of all chopping tools

Ashes

- ❑ Metal pail for cleaning out firebox
- ❑ Tightly covered metal pail for outdoor storage away from combustible materials

place a five-dollar bill across the door's gasketed area, then close and latch the door. If you can pull the bill out easily, your stove's seal needs to be renewed. Try and adjust the door latch first. Some stoves have a mechanism for door adjustments, as gaskets tend to shrink through use. If it's not possible to adjust the door or if, after the adjustment is made, the five-dollar bill can still be pulled out easily in one or more places, you should replace the gasket. When the gasket is compressed too much, no adjustment can re-create a good seal.

Replacing gaskets. The materials used to make gaskets have evolved over the years from asbestos to fiberglass. Gaskets are sold in ropes of various sizes and densities. Typical woodstove gaskets are ⅜ to 1 inch (1–2.5 cm) thick. If in doubt about the proper size and density, remove the door and take it to a wood-heat retailer to test a variety of gaskets in the groove.

To install a new gasket, remove the door and place it flat on a protected surface that won't scratch the door's finish. Remove the existing gasket; with some stoves, you'll need to disassemble the door first. Remove any lumps of old gasket cement from the gasket groove with an old screwdriver, then clean the groove thoroughly with coarse steel wool to produce a clean surface to which new cement will adhere.

Dry-fit the gasket first to measure the length you need; cut it a little bit longer than necessary so that you can tuck the ends into each other to form a good seal. Apply a narrow ¼- to ½-inch-wide (6–13 mm) bead of gasket cement or common silicone sealant along the entire groove. Starting on the longest, straightest part of the groove, lay in the gasket, being careful not to stretch or bunch it. Press the gasket into the cement. Once dry, mount the door and slam it lightly. You should hear only the muffled sound of the gasket hitting the stove body. Test the seal with a five-dollar bill.

Cleaning door glass. Woodstoves with ceramic glass panels that have been built since the mid-1980s typically have an air-wash system that sweeps the combustion air down between the glass and the fire so that soot cannot stick to the glass. When combined with seasoned fuel and good operating techniques, this system keeps the glass door clear even after weeks of daily operation. After a period of use, though, a white or gray haze may form on the glass. When the stove is cold, remove this film by wiping the glass with a small amount of light ash applied on a damp paper towel.

Light brown stains that form in the corners of the glass can be removed with a special woodstove glass cleaner. Dark stains that are difficult to remove are a telltale sign that the stove's air-wash system is inadequate or that the stove is operating at too low a temperature. It could also mean that your wood is too wet. Keep your glass clean by operating the stove according to the manufacturer's instructions and by never letting the fire smolder. Never use abrasives or razor blades to remove stains on door glass. These may work on window glass, but woodstove door glass is softer, and either will almost certainly scratch it.

Touching-up paint. If your stove's painted finish has faded and you'd like to touch it up or repaint it without having to uninstall the stove and transport it to a shop, stove paint that will withstand high temperatures is widely available. It can be purchased in spray cans, and it dries to the touch in about 15 minutes. Apply the paint when the stove is cold, taking care to mask any parts you don't want painted. Protect everything around the stove from stray paint spray by laying down large sheets of cardboard or layers of newspaper.

Cleaning firebrick. The walls and floor of most wood-burning stoves are lined with firebrick to protect the unit's steel or cast iron and to increase firebox temperatures for better combustion. Any firebricks that crack should be replaced, although as long as they remain in position, they do not have to be replaced immediately. Most stoves and furnaces use standard-sized firebricks, called "splits," which are roughly half the size of normal house bricks. Standard splits can be purchased at building-supply stores, but some advanced-technology CSA- or EPA-certified wood heaters use lighter, lower-density bricks for higher performance, and these must be purchased from a woodstove dealer. If you need to replace firebricks, be sure to do so with the same type in order to maintain your stove's efficiency.

Cleaning and replacing catalysts. In catalytic stoves (see "Cats & Non-Cats: Two Different Breeds of Stove" on p. 62), the palladium in the catalyst, which lowers the ignition temperature of the gases produced by the fire and cleanly burns off most of the smoke, deteriorates over time. It typically lasts 12,000 hours, or about six years, if cared for properly and used according to the manufacturer's instructions. After a few years of use—or when you see an unusual change in stove performance—it's time to inspect the catalyst and determine whether it needs to be cleaned or replaced. This inspection can usually be done without removing the catalyst from the stove. If the catalytic element looks sound, with no discoloration or missing pieces, and you haven't noticed a drop in your stove's heat output, the catalyst is probably still functioning. You can also check the condition of the catalyst by watching the smoke at the top of the chimney. As the stove heats up, smoke leaves the chimney, but the smoke should decrease dramatically or disappear completely when the catalyst kicks in. If the smoke does not decrease or disappear, there may be a problem. To clean the catalyst, simply remove it from the stove according to the manufacturer's instructions, and gently vacuum or sweep it with a soft brush. Never scrub the catalyst or use abrasives to clean it. If a catalytic element is missing pieces or if its coating shows signs of peeling or flaking, it is time to replace it.

Chapter 6

Fire

FLAMES, SMOKE AND EMBERS—A WELL-BUILT fire is a careful balance between all three. The red-orange glimmer of dancing flames creating the firelight by which the Old Masters painted their famous still lifes; the delicate scent of fruity smoke evoking many a childhood memory of evenings spent by the fire; glowing, nearly translucent orange embers radiantly beaming their warmth across the room in the wee hours of the morning. The trinity of fire represented by flames, smoke and embers is the living embodiment of what happens when wood burns, unlocking and releasing the sun's energy stored deep within its split, seasoned grain. If you hope to achieve the perfect fire, the first step is gaining an understanding of how this trinity drives the combustion process.

What Happens When Wood Burns?

Anyone who has gazed into a flame has marveled at one of the seeming mysteries of life, yet within its mercurial orange-yellow flicker is a complex, ongoing chemical reaction that fuses fuel, oxygen and heat. Wood requires all three of these to burn and in sufficient quantities, particularly at the start of a fire.

As firewood burns, it goes through three phases of combustion:

Evaporation. Up to half the weight of a freshly cut log is water. If wood is properly seasoned (see "Seasoning Wood" on p. 36), its moisture content has been reduced to less than 20 percent. As it is heated in the firebox, this water evaporates, consuming heat energy in the process. The wetter the wood, the more heat energy is consumed. That is why green firewood hisses and sizzles and is hard to burn, while properly seasoned wood both ignites and burns easily.

Pyrolysis. As wood heats up past the boiling point of water—to around 250 to 300 degrees F (120°C-150°C), its cellulose begins to decompose and emit volatile hydrocarbons and tar droplets and the wood begins to smoke. As the temperature rises beyond 575 degrees F (300°C), the hydrocarbons ignite and combust with oxygen in the surrounding air and the smoke catches fire, creating flames and releasing energy in the form of heat and light. The variations in flame color are caused by fluctuations in temperature. The lower blue part of a flame, where carbon monoxide takes on another atom of oxygen to become carbon dioxide, is the hottest, while the upper yellow-to-orange parts, where soot is burning, is the coolest. The presence of other minerals or metals in wood can create dazzling colors—a stray copper nail produces turquoise flames, while driftwood impregnated with sea salt burns a rich green.

Charcoal. Once most of the hydrocarbons and tars have been vaporized, what remains is charcoal. Made almost entirely of carbon, charcoal burns easily with a red glow and very little flame or smoke. Once hot enough, and as long as there is a steady supply of fuel and oxygen, firewood continues to burn through these three phases.

In any fire, all three phases of wood combustion are happening simultaneously, even within a single log. As wood smoke is flaming and the edges glow red with charcoal, water in the core of the log is still evaporating. The challenge in burning wood cleanly and efficiently is to boil off all that water content quickly and make sure the smoke burns with bright flames before it leaves the firebox. Smoke represents about half the total energy content of the wood you burn—the other half is charcoal. Smoke that does not burn in the firebox either condenses in the chimney as creosote deposits or escapes up the flue as air pollution.

Bottom Up or Top Down?

When it comes to building a fire, there seem to be more variations than there are towns in Tuscany, and enthusiasts will tell you their way is best, citing country lore passed down from their grandfathers to back it up. Yet whether built in an open fireplace or a wood-burning stove, started top down or bottom up or fired from the previous night's coals, any good fire has the same basic needs: It must be kindled to quickly heat up the chimney and firebox and create the conditions for a stable, brightly burning fire; it needs good air flow to promote combustion (a fire needs oxygen just as we do); and above all, it requires careful building and planning.

As every fire and stove is unique, you need to know all the quirks of your wood-burning system inside and out to build the perfect fire. For a fire to ignite quickly and build to full intensity without smoldering, you need a source of ignition, some pre-kindling material, kindling (thinly chopped sticks) and a full basket of seasoned logs, preferably of different sizes and species. If a fire is prepared properly, it should take only a single match to light it.

The Bottom-Up Fire: Step-by-Step

Whether in an open fireplace or a wood-burning stove, the bottom-up method is the traditional way to build a fire. It starts with pre-kindling materials like newspaper, cardboard, fatwood, twigs or fire starters produced with sawdust and wax. When using newspaper for pre-kindling, avoid those with color coatings, as they will inhibit flames; avoid glossy magazines for the same reason. Follow each step carefully, or your fire will fizzle.

1. In an **open fireplace**, build a U-shaped wall of 3- to 4-inch-diameter (7.5–10 cm) logs with cut ends facing inward, and place your

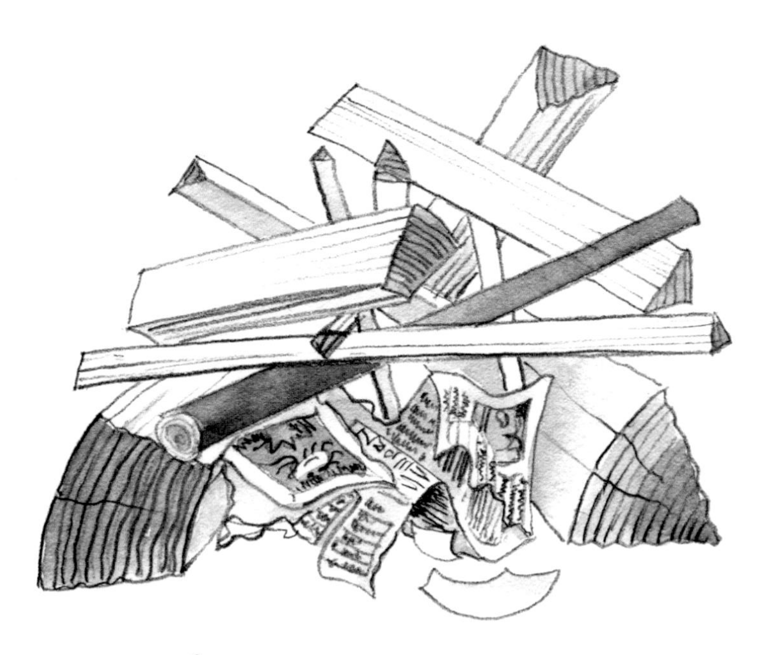

Fig. 6.1. Bottom-up fire

pre-kindling material in the middle. Citrus peel, dried pinecones or bark that has fallen off seasoned logs also make perfect kindling; the oils in some bark provide rich aromatics as well. (Or you can make a V-shaped log wall of thinner wood pieces stacked log-cabin-style to catch quicker—the space between the pieces allows for more air circulation and faster burning.)

2. Lay a lattice of dry, finely split kindling over the pre-kindling material. Straight-grain conifers, like pine, cedar and spruce, make better kindling than do harder woods, as they are easier to split into small pieces and they light more readily. The smaller and drier the kindling, the easier your fire will be to start. Crisscross the kindling so that there is plenty of space between each piece for air circulation; wood that is packed too tightly will not catch. Set one or two layers of crisscrossed pieces of larger wood on top of the kindling. Don't put logs atop the lattice of kindling as the weight will collapse the pile and extinguish your fire.

3. Before you light the fire, make sure the chimney is drafting upward. Open the fireplace damper, and check to see whether there is a cold draft coming down. If there is, your chimney draft has reversed itself. To start the flow of warm air up the chimney, hold a burning piece of newspaper or cardboard above the damper. With some older fireplaces, you may have to open a window in the room to coax air up the chimney.

4. Light the pre-kindling material, and let the upward action of the flames do the work. The early fire is delicate and needs careful tending before it can sustain itself and grow into a roaring blaze. Once the kindling catches, you need to create a sufficient critical mass of heat. If your lattice of kindling is open enough to allow air flow and flames to lick through but closed enough to trap heat, the larger pieces of wood atop the kindling should start to burn.

In a **wood-burning stove**, the process follows the same basic steps. Note: While these steps apply to many wood-burning appliances used in Canada and the United States, they are not intended to replace the manufacturer's instructions for a specific model. Some combustion systems, notably catalytic systems and masonry heaters, require special firing techniques. Before building a fire in a wood-burning appliance, always refer to the instruction manual first.

1. Locate the point where combustion air enters the firebox. For most modern high-efficiency wood-burning stoves and fireplaces with glass doors, air enters the firebox through a narrow strip above and behind the glass panel and reaches the fire at the level of the coal bed. This "air wash" flows down the window and keeps the tarry smoke from blackening the glass. Older stoves without a glass air-wash system have an air inlet toward the bottom of the loading door. With most stoves, the best place to light your fire is just inside the loading door.

2. Build a log wall (see Step 1 for open fireplaces, above), and place your pre-kindling material in the center. The amount of

pre-kindling needed depends on the firebox size and the dryness and size of your kindling. The drier and thinner the kindling, the less pre-kindling you need.

3. Arrange pieces of dry, finely split kindling in a lattice on top of and behind the pre-kindling material. Set slightly larger pieces of split firewood on top of the kindling, and continue to place larger and larger pieces on top—making sure they are supported by the sides of the log wall—until the stove is over two-thirds full. Don't add unsplit logs until the fire has been started and is "bedded in," or fully self-sustaining. Otherwise, your carefully built fire may collapse in on itself.

4. Make sure the air control(s) and bypass damper (if the stove has one) are fully open, and check to see whether your chimney is drafting upward. If it isn't and you feel cold air coming down, place a balled-up piece of newspaper as high up in the stove toward the chimney (usually above the baffle plate) as you can get it, then light it—it should get sucked upward and reverse the chimney draft with its warmth.

5. Light the pre-kindling. Leave the stove's air control(s) and damper fully open; in fact, it often helps to keep the stove door slightly open until the fire catches and spreads through your load of wood. Don't make the mistake of closing the air control(s), damper or door too soon. The flames may look good, but until the stove and chimney have warmed up and a bed of coals is established, the fire is not at the critical mass it requires to sustain itself. You may need to add larger pieces of split firewood until the fire reaches this point. Leave the air control(s), damper and door open until the firebox is full of flame and the wood is charred black with glowing red edges, then close them down in two or three stages.

The Top-Down Fire: Step-by-Step

This seemingly counterintuitive technique—building a fire "upside down" and lighting it from the top—is becoming more popular among woodstove owners. The advantages of this method are minimal start-up smoke, little chance that the fire will collapse and smother itself and, if you are building it in a wood-burning stove, no need to keep opening the loading door to add larger pieces until the fire is bedded in. Building a top-down fire can be tricky in some wood-burning stoves because of the limited height of the firebox, but by keeping the stack of wood fairly low, it can be done. The secret is to make sure the pieces in each successive layer are a little smaller than the ones below.

1. Place three or four full-sized pieces of split, dry firewood on the floor of the firebox or fireplace, evenly spaced and perpendicular to the stove door. If you use wedge-shaped pieces, make sure the wedges are pointing up.

Fig. 6.2. Top-down fire

2. Place a layer of slightly thinner pieces perpendicularly across these logs, leaving enough room for proper air circulation.

3. Place another layer of thinner pieces perpendicularly across the second layer. Build successive layers until the stove is almost two-thirds full.

4. Place a layer of coarsely split kindling perpendicularly across the top.

5. Add a layer of finely split kindling perpendicularly across the thicker kindling.

6. Finally, place pre-kindling on top of the pile. As crumpled sheets of newspaper tend to roll around and fall off the fuel as they burn, try making newspaper "knots": Roll up a full sheet of newspaper from corner to corner or fold it into a 1-inch-wide (2.5 cm) strip, then loop or tie a knot in it. Place four or five newspaper knots on top of the kindling.

7. Check for draft, and if the draft is reversed, follow Step 3 in the instructions for building a bottom-up fire. Set the air control(s) and damper to fully open, light the pre-kindling, and close the loading door.

Flash Fires and Extended Fires

In the context of home heating with wood, a **flash fire** describes a small amount of wood that is burned very quickly. During mild weather, it is a useful technique to avoid overheating the living space and causing the wood to smolder. To build a flash fire, rake any coals into a pile toward the air inlets at the front of the firebox, and load at least three small pieces of firewood on and behind the coals. These pieces should be loosely stacked in a crisscross arrangement. Open the air inlet to produce a bright, hot fire. The air supply can be reduced slightly as the fire progresses but not so much that you extinguish the flames. When only charcoal remains, the air supply can be reduced

Fig. 6.3. Firewood placement for a flash fire

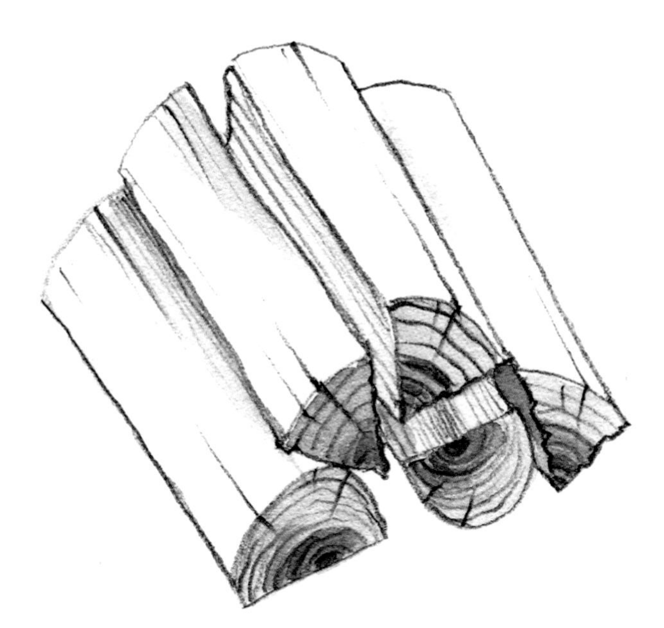

Fig. 6.4. Firewood placement for an extended fire

further to prevent cooling the coal bed. Use flash fires in spring and fall when you want to take the chill off the house.

To achieve a longer-lasting burn, build an **extended fire** by raking the coals toward the air inlets—this will raise the temperature in the firebox and allow the wood to catch fire. Place larger pieces of wood (3 to 4 inches in diameter) compactly against the rear of the firebox. Keeping the pieces close together prevents the heat and flame from penetrating the load; it also preserves the buried pieces for later in the burn cycle. Open the air inlet fully, and leave it open until the firebox is full of flames and the surface of the wood has a thick layer of charcoal and is burning brightly. This will take around 5 to 20 minutes, depending on load size and how seasoned the wood is. Then reduce the air setting in stages so that the flames slow down but, again, not so much that you extinguish them. The charcoal layer insulates the rest of the wood and slows the release of combustible gases. This allows you to turn down the air control and still maintain a clean-burning fire. Use the extended-fire technique to achieve an overnight burn or a fire to last the day.

If you've built your fire correctly, using seasoned wood, the suggested air settings and a proper loading arrangement, a new load of wood should ignite almost instantly. If your wood is burning cleanly, this is what you should see:

- Wood flames until only charcoal remains. If there are no flames, something is wrong.
- If there are firebricks in the firebox, they should be tan in color, never black.
- Steel or cast-iron parts in the firebox are light to dark brown, never black and shiny.
- If the appliance has a glass door with air wash, it is clear.
- If the appliance has a glass door without air wash, it is hazy but never black.
- The exhaust coming from the top of the chimney is clear or white. Blue or gray smoke indicates smoldering, poor combustion and low fire temperatures.

Tending and Maintaining Your Fire

About 15 to 20 minutes into your new fire, the small logs are blazing above the first bright embers on a nest of fine ashes and your log wall is burning nicely. Your goal now is to tend the fire and build up enough embers in a central hot spot to ensure that the fire is hot enough to take any log you feed it.

Open Fireplaces Versus Wood-Burning Stoves

The difference between tending a fire in an open fireplace and tending a fire in a wood-burning stove has to do with exposure to the air. Because the tightly sealed firebox in a high-efficiency wood-burning stove allows for the full combustion of wood smoke, up to half of a log's heat potential can be reached. In an open fire, that's not possible, so managing a bed of embers is the only way to ensure a good fire that radiates its warmth into the living space.

To create good embers, firewood must be as dry as possible. Building a U-shaped log wall (see "The Bottom-Up Fire: Step-By-Step" on p. 117) will produce a wall of glowing embers as the split faces of the wood char and burn, radiating heat outward. At this point, move half-burned logs toward the center and feed new logs to the left or right of this hot, brightly burning pile, with the cut surface facing inward.

Dense hardwoods, such as oak, maple and hornbeam, guarantee good embers. Make sure your firewood is at room temperature. Throwing ice-cold logs onto a bed of embers is like turning a garden hose on your fire—they will immediately cool down the fire and may even put it out. Your logs should be halved or quartered so that they cannot roll out of the fire. If you have an open fireplace, invest in a fireplace grate or a pair of andirons (see "Hearth Accessories" on p. 132) to prevent burning logs from migrating beyond the hearth.

If the aim of your fire is not so much about generating warmth as creating flames and firelight to cast an exquisite glow and dancing shadows around the room, choose small-diameter logs of denser and more brightly burning species, such as birch, beech, oak or the hard maples. Build your fire with slightly wider gaps to promote more flames—you will have to feed it more often, but it will provide a pleasing display of vivid flame. Whether you build a fire for flames or warmth, resist the temptation to poke it to release a shower of bright sparks and watch them float up the chimney. Poking a fire repeatedly thins out the fire and causes it to burn out more quickly.

Maintaining a fire in a wood-burning stove is an easier affair. Rather than managing embers to provide maximum heat, as in an open fireplace, you are primarily managing air flow. Fresh logs are simply fed into the stove when needed. Logs cannot roll out of the stove when the door is closed, and the issue of sparking wood is avoided altogether.

Once you start feeding the main logs to the stove, it pays to be generous. Too small a fire at this stage will not heat the space efficiently. Avoid loading only one or two pieces of firewood at a time on a coal bed; a minimum of three are needed to form a sheltered pocket of glowing coals to sustain the fire. When the fire is "bedded in," you can slow down and feed one or two large logs at a time when the previous ones have burned to embers. This will maintain an even heat and maximize your stockpile of wood.

A Clean or Dirty Burn?

At this point, maintaining the fire is a delicate balance between the cleanest possible burn and a burn that provides you with the greatest amount of usable heat. Too far in the former direction leaves you with a squeaky clean chimney, but you'll be burning a lot of wood; too far in the other direction saves wood but creates dangerously dirty stovepipes and chimneys. Indeed, how you stoke and maintain your fires determines how quickly you will burn through your woodpile and how effectively you heat your house.

The key to striking this balance lies in stoking moderately, regulating the temperature and firing in cycles.

Stoke moderately. After kindling, never add more wood than the fire can handle. Instead of dropping in a large unsplit log, reload with fairly small split pieces of wood when most of the previous load has burned down to large, glowing coals. Rake the coals toward the air inlet before adding the new wood.

Regulate the temperature. The temperature of a fire is largely dictated by how much oxygen is feeding it. For example, in the pyrolysis stage (see "What Happens When Wood Burns?" on p. 115), the fire throws a serious amount of heat. Continuing to feed the fire oxygen at this stage causes it to burn out rapidly, but restricting the oxygen too much prevents the smoke from being burned off. Flue-gas temperatures below 400 degrees F (200°C) indicate that more oxygen is needed, while temperatures in excess of 900 degrees F (480°C) can potentially damage your chimney. For older wood-burning stoves without glass doors, a stovepipe thermometer can help you determine whether your fire is hot enough to burn cleanly. For open fireplaces and modern wood-burning appliances with glass doors, simply watch the fire. If your fire is not smoldering and the wood flames brightly until it is reduced to charcoal, you are burning cleanly. Light gray, almost white smoke issuing from your chimney is also a positive sign.

Fire in cycles. Wood fires burn best in cycles. A cycle starts with the ignition of a new load of wood and ends when that load has been reduced to a glowing coal bed. Each cycle should provide between three and eight hours of heating, depending on how much wood is used and how much heat is needed. Plan the firing cycles around your household routine. For example, if someone is home all day, two four-hour fires allow better control of house temperature than one eight-hour burn. Never fully stock a fire and let it go nearly out to get the most mileage out of your wood supply. Your chimney cools down between firing cycles, which increases the possibility of creosote buildup. If your chimney is an exterior one, that's virtually guaranteed.

Controlling Heat Output

To control the amount of heat thrown by your fire, try the following techniques:

Wood species. In mild weather, burn light woods, such as poplar, aspen, pine and willow. When properly seasoned, these woods light fast, burn quickly and don't leave a big coal bed that can overheat the room the way hardwoods can.

Cooking on a Wood-Burning Cookstove

As any Scout will tell you, the culinary possibilities of cooking over an open flame are somewhat limited. A wood-burning cookstove, however, offers the opportunity to cook food the way our great-grandparents did: on a cast-iron or porcelain stovetop and in the oven.

Indeed, mastering cooking on a wood-burning cookstove is one of the greatest satisfactions of the wood-burning lifestyle. It is not without its challenges, however, the most obvious of which is maintaining an even temperature without a thermostat. When you put something in a wood-burning oven to cook, the temperature drops as the food absorbs the heat and the stove cools down. You need to feed more wood to the fire to bring the temperature back up, and knowing exactly how much wood and when to add it can be determined only through trial and error—and likely a few burned muffins.

Cooking on the stovetop is an easier place to start. Basic frying, slow-simmer soups, chilies and stews aren't too difficult, and there's the additional benefit of their being flavored with the aromas of the wood being burned. Hickory and apple wood are tried-and-true favorites. Keep a kettle, canner or pot of water on your cookstove at all times so that you'll have low-boiling water at hand for adding to soups and stews, washing dishes or making a cup of tea. As the surface of most cookstoves get hotter than the boiling point of water, you will need to use trivets—ornamental cast-iron stands with short legs—to effectively simmer. Never put a pot of cold water directly on a hot stovetop, as this can lead to rusting, pitting or actually cracking the stovetop. Take extra care not to let pots spill over onto a hot stovetop, as the food will rapidly burn and fill your house with the tang of scorched food for days.

Pots and pans should be either ceramic or solid cast iron, with lids, carry hooks and lifters. (Stainless-steel and aluminum pans will blacken quickly and can become distorted from the cookstove's intense heat.) Dutch oven pots and heavy pans are best for the stovetop. In the oven itself, use heavy roasting pans and cake tins. Before using, season your pots and pans by coating them with oil and heating them to a smoking temperature.

Each wood-burning cookstove is different, so learning how to successfully cook on one requires getting to know your stove's idiosyncrasies and "hot spots." Experiment with the dampers to increase or decrease the intensity of the fire. Familiarize yourself with different wood types—fast-burning softwoods to cook something hot and fast; harder woods to burn slower and longer for slow-cooking stews and breads.

To maintain an even baking temperature requires experience, constant vigilance and a ready supply

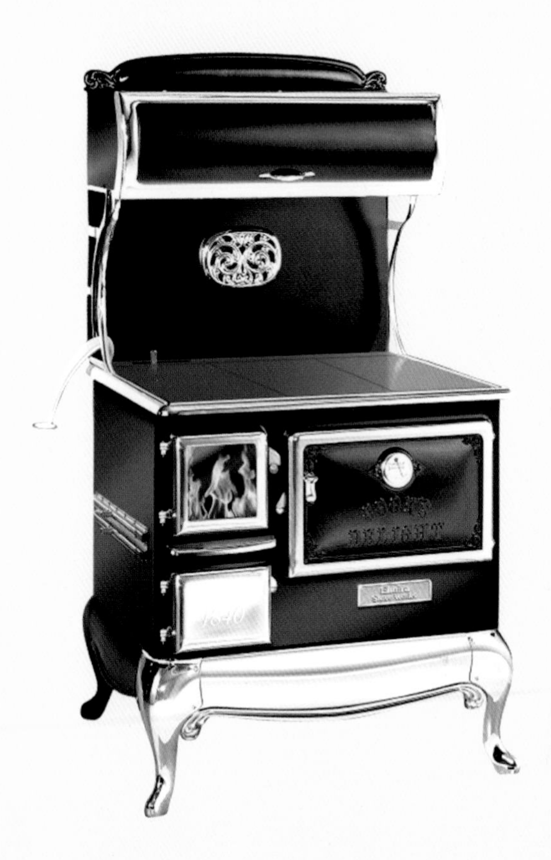

of firewood. Because very few old woodstove ovens had temperature gauges, most traditional recipes do not specify exact cooking temperatures and times but simply suggest using a cooking style that's "slow," "moderate" or "fast." Here are a few tips for figuring out the temperature of a wood-burning stove without an oven thermostat:

- Throw some flour onto a baking tray, and put it in the oven for a few seconds. If it starts to change color, the oven is warm; if it quickly turns brown, the oven is very hot.
- A piece of writing paper placed in the oven curls up and browns when the stove is at the proper heat for baking pastry.
- If a drop of water sizzles and dances on the stovetop, the stovetop is pretty much ready to cook on.
- If you can't hold your hand flat a couple of inches above the cooking surface for more than three seconds, it's hot enough to cook on.

Load size. Build small fires in mild weather and large loads in winter.

Load configuration. In mild weather, load the wood crisscross for fast, hot, brightly burning fires. In cold weather, load the wood compactly to the rear of the firebox to build longer, higher-output fires (see "Flash Fires and Extended Fires" on p. 122).

Load orientation. If your firebox is square, load the wood in either an east-west or a north-south formation. Use the east-west formation in mild weather or for overnight burning to achieve an extended low-heat output—wood breaks down more slowly when the combustion air reaches it from the sides. Load the wood north-south in cold weather, because the heat and flame can penetrate the load more easily, producing a sustained high-heat output.

Overnight Burning

While the idea of keeping a fire going all night may seem romantic, it is not recommended for safety reasons, particularly with an open fireplace, where there's a chance that a burning log may escape the hearth and roll onto the floor. In a wood-burning stove, the price you pay for a good night's sleep without having to tend the fire may be a chimney full of creosote.

If you must keep a fire going in an open fireplace for reasons of heat, put a large log on the embers two to three hours before turning in. Use andirons to hold the log in position, and make sure you have a fireguard, or metal screen, in place. The log should burn slowly and steadily throughout the night. In the morning, the underside of the log will be crisscrossed with glowing embers that can be used to kindle a new fire. Another way of keeping a fire from going out is to build up a good quantity of embers and cover them with about 1 inch (2.5 cm) of white ashes from the edge of the fire. These embers can be dusted off in the morning and combined with kindling to bring the fire back to life.

When using a wood-burning stove overnight, follow the manufacturer's instructions, stacking the wood until the firebox is nearly full. To prevent the embers from burning away, close down the air control(s) and damper to allow in just a trickle of air. It will take some practice to find the right setting. Your goal is to have a bed of coals eight hours later, enough to start a new fire. Never let the fire smolder. As long as there is solid wood in the firebox, there should be active flames. With an advanced-technology, medium-sized stove, it is possible to achieve a reliable overnight burn while maintaining flaming combustion and still creating enough charcoal in the morning to rekindle a new load. If you heat with wood around the clock, do not get into the habit of "burning the stove off" in the morning by firing a very hot cycle for 15 to 20 minutes to heat up a cold stove. This is the perfect recipe for a chimney fire, as the intense temperatures may ignite any creosote lining the chimney.

Charcoal and Ashes

Once upon a time, in an age before fire starters and butane lighters, kindling a fire was a tremendous challenge, particularly in cold northern climates. As a result, fires were kindled and kept alive continuously, with the new day's fire ignited from the warm embers of the night before. If you are deeply invested in the wood-burning lifestyle or simply heating your home with wood in the middle of a bitter winter, you will likely want to try your hand at this time-honored tradition.

Charcoal

In a wood-burning stove, look for any live coals that remain at the back of the firebox after the fire has burned down. Remove a small amount of ash from the front, and carefully rake the live coals to just inside the front of the loading door, where the combustion air enters. Place at least three to five pieces of firewood on and behind the coals. Always place the smallest, driest piece of firewood directly on the coals to act as the igniter. If you have very few coals, you may need to add some kindling.

Open the air control(s) fully, and close the door. Your new load of firewood should ignite quickly. Leave the air control(s) open until the firebox is full of bright flames and the wood is charred black, anywhere from 5 to 20 minutes. When the wood is charred, reduce the

Hearth Accessories

To efficiently manage your wood-burning fire and keep your hearth tidy, invest in a set of specialized tools. Here are some of the most common:

1. **Andirons.** Also known as firedogs, these ornate horizontal bars were originally designed to support burning firewood, keeping it off the fireplace floor and aiding in air circulation. Made of iron, steel, copper, bronze or silver, they once ranged from the simple to the elaborately ornamental but are now used more for decoration. Some wood burners maintain that andirons are not as safe as a concave fireplace grate, which cups the wood and prevents it from rolling out of the fireplace, but a grate is nowhere near as elegant.

2. **Ash container.** Ashes should be removed regularly for safety and to make room for new fires. If your wood-burning appliance doesn't have an ash pan, you will need a small metal ash bucket to transfer the ashes to an outside ash-storage can, as the ashes may hide dormant coals and release toxic carbon monoxide into the indoor air (see "Ashes" on p. 134). The ash container in the illustration on p. 133 is doing double duty as a storage unit for kindling.

3. **Bellows.** If you have built your fire properly (see "Bottom Up or Top Down?" on p. 117), you may not need bellows, but when a fire is beginning to die out, using bellows to force a steady stream of air to revive the fire beats kneeling down and blowing. Some bellows have intricate inlay and designs that rival a luthier's work. The bigger the bellows, the faster your fire recovers.

4. **Fireback.** Once designed to protect the rear wall of the fireplace from fire damage, this decorative, heavy, cast-iron plate improves the heat output of your fireplace by absorbing warmth from the fire and reflecting it back out into the room.

5. **Fireplace screen.** Designed to cover the fireplace opening completely, a fireplace screen keeps sparks, embers and logs in while keeping pets and children out. It is available in wrought-iron, mesh or glass designs, and some models have doors. Spark-guard curtains made of stainless-steel mesh are another option. (A screen is not recommended for use with a wood-burning appliance as it allows smoke to spill into the room and leads to lower efficiency, as much of the energy in the firewood is wasted with the door open.)

6. **Fire tool set.** Tool sets for wood-burning appliances are

Fig. 6.5. Fireplace gear

simpler and shorter in length than a standard set for an open fireplace. Available in an array of plated finishes, a basic fire tool set should have a rake for moving charcoal and logs around, a brush and a small shovel.

7. **Firewood holder.** These come in a wide variety of designs and sizes, from Shaker to Bauhaus. Some have compartments to stack different sizes of split firewood. Look for a metal or an iron model with an open design that allows your firewood to continue drying.

8. **Hearth gloves.** A pair of heat-resistant gloves is handy when you need to rake a large coal bed or, in an emergency, pick up a log that has rolled out of the fireplace. Specialized hearth gloves are thick, lined leather gloves with long cuffs. If there isn't a wood-heat dealer nearby, a welding supply store carries gloves that will do the trick.

air setting in stages to produce the amount of heat and length of burn you desire. If the wood stops flaming, open the air control(s) again, and let the fire burn longer, until the firebox heats up fully.

Some non-catalytic stoves (see "Cats & Non-Cats: Two Different Breeds of Stove" on p. 62), with their insulated fireboxes and high temperatures, do such a good job of burning off the wood smoke that large amounts of charcoal are produced. In moderate weather, this coal bed can provide adequate heat, but as temperatures plummet and a fire is needed, a large coal bed tends to get in the way of your adding more wood. A quick solution is to rake the coals toward the stove's primary air inlet, place one smallish log atop the pile, and open the air control(s) and damper wide. The log will ignite and burn quickly, not only consuming the coal bed but also generating tremendous heat. Repeat two or three times, if necessary, to reduce the size of the coal bed.

If you are using warm coals to rekindle a fire in an open fireplace, carefully make a shallow bed in the ashes about 12 inches (30 cm) wide, rake the coals into the middle of the ashes, and either use bellows or gently blow on the coals until they glow brightly. Place pieces of pre-kindling (twigs, bark, citrus peel, cardboard, newspaper knots) around them, leaving the side facing you open. Blow on the embers until the pre-kindling catches fire. If the result is a lot of smoke, blow slightly harder to raise the heat of the embers and encourage the smoke to catch fire.

Ashes

Like any solid fuel, wood leaves a residue behind once it is burned. When a fire dies down or goes out, ashes blanket the floor of the fireplace or the firebox. Many former wood burners claim that cleaning out the ashes is one of wood heating's biggest drawbacks. In reality, when firewood is burned properly, less than 1 percent remains as ash, and that ash can actually prove useful in creating a bed for your next fire.

Over the years, wood ashes have been put to many practical uses. Containing between 25 and 45 percent calcium carbonate, up to 10 percent potash and about 1 percent phosphate and minerals, wood ashes make an excellent dry fertilizer and nutrient mix for gardens in areas with high rainfall, where soils tend to be more acidic. Wood ashes have traditionally been used to make soap and pottery glazes and to leach the bitterness from olives. Mixed with water into a paste, wood ashes can be used to clean silver and glass, including the glass window in your woodstove.

Nevertheless, ashes must be removed frequently. Your fireplace or wood-burning appliance works best when you remove a small amount regularly rather than letting it build up over several days. To make raking coals and kindling loads easier during round-the-clock heating in cold weather, remove a small amount of ashes each morning before the new fire is kindled. Many advanced- technology wood-burning stoves feature ash pans below the firebox for easy removal. Never let ashes build up to more than 2 inches (5 cm).

Don't wait until the fire is cold to remove ashes. When they are still warm, the chimney draft helps draw any displaced dust back into the stove or fireplace. The best time to remove ashes is first thing in the morning, while there is still live charcoal at the back of the firebox and enough heat to produce some draft.

Use a hand shovel or gardening trowel to transfer ashes into a small metal pail or container, and move them outside a safe distance from the house, as they release carbon monoxide. Always use a metal container, as live embers can stay dormant for days insulated by the surrounding ashes. Never store the ashes in the same pail you used to remove them; an open pail of ashes on a wooden deck outside needs only a breeze to stir the contents and expose the embers to ignite a deadly house fire. Instead, store ashes safely outside on a stone or concrete surface or on the ground in a metal container with a tight-fitting lid and recessed bottom. When completely cold, the ashes can be used to fertilize your garden or disposed of with the compost.

In many ways, ashes are just as alive as the fire that produced them—they should be treated with the same respect and care. Never leave them unattended for long.

Chapter 7

Product Guide

THE FOLLOWING WOOD-BURNING APPLIances are representative examples of what is currently offered by North American wood-heat retailers, from simple woodstoves to high-end masonry heaters and cookstoves. All are available in the United States and Canada. When shopping for a wood-burning appliance, be sure to do as much homework as possible—visit several wood-heat retailers, look for professional certifications like Wood Energy Technology Transfer (WETT), the Chimney Safety Institute of America (CSIA) or the Masonry Heater Association of North America (MHA), familiarize yourself with local regulations and building codes and ask the advice of professional contractors. As indicated, all prices are in U.S. dollars and reflect prices shown on the company's websites in 2014. Taxes and any other charges have not been included.

Wood-Burning Stoves

Lennox Striker S160

www.lennoxhearthproducts.com

This compact EPA Phase 2-certified stove is one of the cleanest-burning non-catalytic woodstoves on the market and is ideal for heating small spaces. Constructed from premium materials, like heavy-gauge stainless steel with a ceramic-fiber firebox blanket that utilizes technology developed by NASA (with 10 times the insulation value of firebrick), the Striker S160 burns wood efficiently, resulting in lower emissions and lower fuel costs.

- Extra-wide viewing glass with integral air-wash system
- Available with pedestal or legs
- Can be customized with a range of arched or traditional nickel-plated doors

Regency Classic F5100

www.regency-fire.com

The Classic F5100 claims to be the largest EPA-certified woodstove available in North America. With its sleek cast-iron design and generous 4.42-cubic-foot firebox, it's capable of delivering up to 80,000 BTU of heat. Thanks to Regency's Eco-Boost technology, this highly efficient catalytic woodstove is capable of burning a 90-pound wood load for over 30 hours with emissions of just over 1.4 grams per hour (g/h).

- 180-square-inch glass viewing area offers an exceptional view of fire
- 4.42-cubic-foot firebox allows 22-inch logs to be loaded front to back and side to side
- Eco-Boost design maximizes re-burn of gases prior to entering the catalytic combustor, prolonging its life

Quadra-Fire Yosemite

www.quadrafire.com

This Washington State-based manufacturer draws its name from its unique Four-Point Burn System, which burns gases in four separate combustion cycles to increase efficiency, reduce emissions and create

Table 7.1

	Lennox Striker S160	Regency Classic F5100	Quadra-Fire Yosemite	Vermont Castings Defiant
BTUs (Output)	68,000	80,000	42,500	75,000
Dimensions (H/W/D)	31"x23"x19"	39"x29"x29"	25"x26"x25"	29"x33"x24"
Combustion Mode	Non-Catalytic	Catalytic	Catalytic	Both
Area Heated†	700–1,400 sq ft	Up to 2,250 sq ft	800–2,200 sq ft	Up to 2,400 sq ft
Burn Time	8 hours	30 hours	9 hours	14 hours
EPA-Certified Emissions	1.6 g/h	1.4 g/h	2.7 g/h	1.1 g/h catalytic; 2.3 g/h non-catalytic
Efficiency	85%	78%	80%	84% catalytic; 76% non-catalytic
Cost	US$1,360	US$3,600	US$1,300	US$3,000

†Figures may vary with individual conditions, such as floor-plan layout, insulation value/heat loss of the house and geographical location.

a beautiful rolling fire. The Yosemite catalytic woodstove uses an elegant and superior cast-iron design to radiate warmth and accent any room.

- Matte black or porcelain mahogany finishes
- Additional side-load door
- Welded steel firebox

Vermont Castings Defiant

www.vermontcastings.com

Made from 100 percent recycled materials with production processes that use renewable energy, this unusual handcrafted cast-iron hybrid woodstove allows you to convert from catalytic to non-catalytic operation in under a minute. Whether you need an efficient catalytic primary heating source or a low-maintenance non-catalytic solution for supplementary heat or ambience, the Defiant fits the bill.

- Exceeds EPA emission standards
- Accommodates 24-inch logs
- Top-load design

Pellet Stoves

Harman P-68

www.harmanstoves.com

Harman's PelletPro technology allows its stoves to burn any grade of pellet with even heat and precision control. Harman offers a line of stoves featuring traditional cast-iron doors, clean lines and customizable decorative slate tiles. The P-68 is a pellet-burning powerhouse, capable of heating up to 3,900 square feet. Basic maintenance can also be performed without shutting down the unit.

- 76-pound hopper and large ash pan make maintenance easier and less frequent
- Exhaust-sensing technology constantly monitors and adjusts heat levels within one degree
- Optional outside-air kit available

Nu-Tec Upland 207

www.nutec-castings.com

Designed to last a lifetime, the traditional cast-iron body of this pellet stove is based on the popular Upland airtight woodstove first produced in the United States in the 1980s. The Upland 207 efficiently radiates comfortable heat without the need for noisy, air-drying convection fans. As a bonus, the heavy top can double as a cooktop and is large enough to hold a steamer, an earthenware dish or a soup pot.

- Available in six porcelain-enameled colors
- Hopper built of heavy-gauge steel
- Can also burn a corn/pellet mix of up to 80 percent corn

Lennox Montage 32FS

www.lennoxhearthproducts.com

Available in four trim kits—Arch, Artisan, Contemporary and Gothic Arch—the EPA-certified Montage 32FS, by Lennox Hearth Products, can be customized to fit any living room and burns both standard- and premium-grade fuel to deliver dependable, low-maintenance heat. Its reflective ceramic glass warmly radiates heat and blocks the firebox view when not in use.

Table 7.2

	Harman P-68	Nu-Tec Upland 207	Lennox Montage 32FS	Wittus Toba
BTUs (Output)	68,000	30,000	32,000	34,000
Dimensions (H/W/D)	36"x23"x29"	29"x29"x28"	28"x22"x24"	43"x20"x20"
Hopper Capacity	76 lb	75 lb	50 lb	50 lb
Area Heated	1,500–3,900 sq ft	Up to 1,500 sq ft	800–1,900 sq ft	Up to 1,800 sq ft
Burn Time	72 hours	72 hours	37 hours	40 hours
EPA-Certified*	(exempt)	(exempt)	Yes	Yes
Efficiency	—	83%	100%	94%
Cost	US$3,900	US$2,599	US$4,499	US$4,900

*Although pellet stoves do not require CSA or EPA certification for particulate emissions to be sold in Canada or the United States, some manufacturers have tested their products for certification.

- Patented UltraGrate™ burn grate achieves nearly 100 percent combustion efficiency
- Easily accessible ash drawer can be removed quickly for simple cleaning

Wittus Toba

www.wittus.com

This tall, sleek, modern pellet stove incorporates state-of-the-art technology and award-winning Italian design. Minimal and striking, the Toba features handmade heat-retaining panels in ceramic, soapstone or sandstone. Unlike most pellet stoves, the heat vents from the top of this unit rather than the front.

- Programmable climate-control system allows up to 60 daily and 10 weekly programs
- Self-cleaning firebox lined with white refractory Alutec

Cookstoves

Elmira Fireview

www.elmirastoveworks.com

This Canadian manufacturer built its name in the 1970s by reproducing the turn-of-the-century Findlay Oval cookstove. Today, its popular Fireview model is a high-efficiency, airtight version of a classic homestead design, with an overhead warmer, nickel-finish accents and claw feet. The spacious 3-cubic-foot oven can accommodate a full-sized roasting pan, and its generous firebox can be loaded from the front or the top.

- Comes in three models, one with a right-side warmer and another dual gas/wood stove with two gas burners
- 48-inch models have a larger cooktop and a side warmer
- Optional water jacket can provide domestic hot water at approximately 10 gallons per hour

Heartland Artisan

www.heartlandapp.com

Solid and rugged, Heartland's Artisan cookstove is a cast-iron marvel, one that borrows its classic design from the early 1900s. The stove's gentle radiant heat is transferred through an internal cast-iron structure, spreading the heat evenly to two ovens and a hot plate, which can be uncovered to send glorious warmth into the room. When food is bathed in heat from all directions, flavors are sealed in without drying. Its hot plate is a single piece of cast iron, making temperature regulation by sliding pots and pans between hot spots a snap.

- Large firebox provides longer burn times and can accommodate 16-inch pieces of firewood
- Temperature gauge on the main oven shows approximate heat levels
- Solid cast-iron construction allows for more radiant heat and lower clearances

Esse Ironheart

www.esse.com

In business longer than any other stove manufacturer in England, Esse opened its doors in 1854, and its traditional designs have served

Table 7.3

	Elmira Fireview	Heartland Artisan	Esse Ironheart
BTUs (Output)	60,000	45,000	36,000
Dimensions (H/W/D)	67"x50"x37"	36"x37"x24"	35"x35"x24"
Area Heated	800–1,800 sq ft	1,200–1,700 sq ft	800–1,200 sq ft
Cooking Surface	50"x29"	34"x25"	35"x24"
Oven size	3.0 cu ft	2 x 1.5 cu ft	1.7 cu ft
Cost	US$5,195	US$8,500	US$6,096

both Florence Nightingale and polar explorer Robert Scott. Designed to mark the 150th anniversary of the company, the Ironheart combines the homestead charm of a wood-burning stove with a functional oven and cooktop in a thoroughly modern cookstove.

- Up to 82 percent efficient
- Primary and secondary air-flow controls allow for the fine temperature modulation necessary for cooking
- Machined hot-plate surface and stainless-steel hot-plate covers
- Boiler model is capable of running a single radiator and providing domestic hot water

Fireplace Inserts

Napoleon 1101

www.napoleonfireplaces.com

Napoleon Fireplaces' 1101 high-tech, non-catalytic wood-burning insert won industry awards for its unique design, which allows it to be mounted in either a masonry or a factory-built fireplace. The innovative heat-circulating blower and thermostatic sensor transform small amounts of wood into large amounts of cozy heat.

- Flush-mount or hearth-mount installations available
- Optional cast-iron surround kit
- Fully refractory brick-lined firebox maintains interior temperatures necessary to achieve low emissions and complete combustion

Harman 300i

www.harmanstoves.com

This modestly sized 300i fireplace insert packs a wallop, generating up to 75,000 BTUs. Harman's FireDome™ technology uses two combustion cycles to completely burn wood, smoke and exhaust, while twin 120 CFM blowers help distribute heat throughout your home. It also has the lowest EPA-certified emission levels for inserts on the market.

- Easily converts to an open fireplace with the addition of an optional fire screen
- Up to 17 hours of steady, even heat from a single load of wood
- EPA-certified emissions 1.1 g/h, 0.8 g/h while on low

Quadra-Fire Voyageur Grand

www.quadrafire.com

The Voyageur Grand is a flush-front fireplace insert with elegant lines and a large, unobstructed viewing area. It harnesses Quadra-Fire's proven Automatic Combustion Control technology, which feeds the fire with air when it's needed most—for precision start-up, consistent, long, clean burns and powerful heat production. Its heavy-duty, cast-iron construction makes the Voyageur Grand a rugged heating option.

Table 7.4

	Napoleon 1101	Harman 300i	Quadra-Fire Voyageur Grand
BTUs (Output)	42,200	75,000	52,200
Dimensions (H/W/D)	22"x29"x16"	28"x41"x17"	23"x33"x18"
Firebox Capacity	1.7 cu ft	3.0 cu ft	2.35 cu ft
Area Heated†	600–1,600 sq ft	1,600–4,300 sq ft	1,100–2,800 sq ft
Burn Time	7 hours	17 hours	14 hours
EPA-Certified Emissions	4.1 g/h	1.1 g/h	3.1 g/h
Cost	US$1,900	US$2,200	US$3,000

- Matte black or porcelain mahogany finishes
- Up to 80 percent efficient in CSA B415.1 testing
- Additional side-load door

Masonry Heaters

AlbieCore

www.mainewoodheat.com

Like most masonry heater cores, the Maine Wood Company's AlbieCore is sold as a unit requiring assembly, with all the actual masonry surrounding it (bricks, stucco or natural stone) to be bought separately and installed by a licensed professional mason. Based on the traditional Finnish contraflow design, the AlbieCore can be installed in a variety of wall-built or freestanding masonry heater configurations.

- 12-hour firing cycle
- Wide range of Finnish cast-iron doors and hardware available
- Creates radiant warmth that does not dry out the air indoors

Envirotech Classic

www.nymasonryheater.com

Empire Masonry Heaters' Envirotech Classic provides the core materials for a finished masonry heater. Like many of the company's designs, this masonry heater has been tested for efficiency and emissions at a third-party laboratory. In 2007 tests at an EPA-certified lab, the Envirotech Classic released only 0.82 gram of particulates per 1,000 grams of wood burned, well under EPA guidelines, setting the bar for masonry heaters and fireplaces alike.

- 12-hour firing cycle
- Soapstone, granite, brick or natural stone finishes
- Optional see-through oven and firebox

Crossfire CF-2100

www.crossfirefireplaces.com

The Crossfire CF-2100 is an all-in-one masonry heater that comes with everything you need to build your hearth: a complete set of modular refractory blocks, steel butterfly damper, steel and glass door, door fasteners and a 55-pound pail of cement.

- Can reach an efficiency of 90 percent with properly seasoned wood
- Ideal for building "top-down" fires

Table 7.5

	AlbieCore	Envirotech Classic	Crossfire CF-2100
BTUs (Output)	dependent on masonry	37,000	dependent on masonry
Dimensions (H/W/D)	48"Wx36"D (core only)	74"x60"x23"	46"Wx16"D (core only)
Area Heated	dependent on masonry	Up to 3,300 sq ft	dependent on masonry
EPA Certified	No	Yes	No
Cost	US$1,900 (core only)	US$7,900	US$4,000 (core only)

- Glass air-wash system ensures a clean and maintenance-free viewing area
- Optional baking ovens available

Wood-Burning Furnaces

Norseman VG2500

www.vogelzang.com

Vogelzang's Norseman VG2500 is an add-on wood-burning furnace designed to be used in conjunction with an existing furnace. Its twin 550 CFM blowers provide up to 115,000 BTUs, heating 2,500 square feet. Its large, firebrick-lined, 7-gauge plate-steel firebox holds enough fuel for several hours of heating per load and is surrounded by powder-coated steel, giving the unit an eye-appealing look. Connection to existing HVAC ductwork is easy with the two 8-inch-diameter heat outlets.

- Large 12"x12"x7" feed door for easy loading
- Accepts firewood up to 24 inches in length
- Adjustable cast-iron spin draft for additional burn control
- Convenient ash door and large full-length ash tray for easy removal and disposal

PSG Caddy

www.psg-distribution.com

The Canadian-made PSG Caddy provides exceptional efficiency due to its unique heat-exchanger system. Before reaching the main smoke pipe, hot gases wind their way around the baffle in the combustion chamber and then move into cylindrical ducts above that serve as the exchangers. The resulting heat, which normally dissipates directly into the chimney, circulates inside the furnace. The Caddy's powerful fan then extracts and pushes all this heat into heating ducts throughout the house.

- Meets both EPA- and CSA-emission standards
- Glass door with cast-iron frame allows view of fire
- Easy to clean and maintain

Newmac WFA-85

www.newmacfurnaces.com

Built under strict ISO 9001 standards, Newmac wood-burning furnaces feature baked-enamel finishes, stainless-steel secondary air systems for efficiency and reliability and some of the best warranties

Table 7.6

	Norseman VG2500	PSG Caddy	Newmac WFA-85	Blaze King Apex CBT
BTUs (Output)	115,000	106,400	85,000	80,000
Dimensions (H/W/D)	41"x24"x50"	48"x26"x30"	43"x26"x54"	40"x27"x48"
Add-On	Yes	Yes	No	Yes
Maximum Firewood Length	24"	22"	19"	26"
Firebox Size (H/W/D)	26"x20"x37"	16"x17"x23"	18"x14"x20"	24"x16"x27"
Efficiency	—	76%	73%	90.7%
Emission Standard	EPA-exempt	EPA/CSA B415.1	CSA B415.1	CSA B415.1
Safety Tests Standard	UL391	CSA B366.1; UL 391	CSA B366.1; UL 391	CSA B366.1; UL 391
Cost	US$1,600	US$3,299	US$3,400	US$4,100

and customer support in the business. The stand-alone WFA-85 offers outstanding performance.

- Stainless-steel secondary air system forces superheated air into the firebox, igniting wood gases and increasing efficiency
- Stainless-steel baffle keeps firebox temperature high to burn particulates and gases, resulting in a cleaner burn and very low emission rates
- With the proper duct design, can be used as a gravity furnace when power is out

Blaze King Apex CBT

www.blazeking.com

This catalytic indoor furnace was one of the first to meet and exceed the requirements of the rigid CSA B415.1 test standard, which measures both overall efficiency and emissions. It can be used with an additional blower as a stand-alone furnace or in combination with an existing furnace system. By combining a high-efficiency secondary-burn system with a catalytic combustor, the Apex CBT dramatically increases the efficiency of the furnace and reduces the emissions to the lowest levels possible.

- Large, 12-gauge, 5.7-cubic-foot firebox can hold up to 120 pounds of wood
- Allows long, overnight burns (up to 14 hours)
- Accepts firewood up to 26 inches in length

Outdoor Boilers

Econoburn EBW-150-O

www.econoburn.com

Designed with help from the New York State Energy Research & Development Authority, Econoburn's outdoor wood boilers burn at temperatures that far exceed typical outdoor boilers—greater than 2,000 degrees F—which burns firewood so cleanly, these boilers produce virtually no creosote, ash or nuisance smoke common with these outdoor fixtures. The EBW-150-O burns half the wood of a normal wood-burning furnace and can be easily installed in conjunction with your existing home-heating system to provide the utmost in convenience, fuel efficiency and savings.

- 25-year warranty
- Burn times of 6 to 8 hours
- Built of 1/4-inch ASME Grade 36 carbon steel in ISO 9001 facilities

Central Boiler E-Classic 1450

www.centralboiler.com

Industry leader Central Boiler's E-Classic 1450 uses a three-stage combustion process to burn wood so completely that combustion efficiencies can exceed 90 percent. With its self-cleaning Xtract™ heat exchanger, the E-Classic produces extremely low emissions per BTU of heat. Using water-to-air or water-to-water heat exchangers, the heat can be conveyed into your home's forced-air furnace, radiant baseboard or radiant floor heating system.

- Triple-insulated fireplace door
- Overlapped all-weather powder-coated steel siding for tight weather seal
- FireStar II app for laptops, smartphones and tablets allows remote control of the boiler over the Internet

Portage & Main Optimizer 250

www.portageandmainboilers.com

This Canadian company has a reputation for some of the most efficient advanced-technology outdoor boilers on the market. The

Table 7.7

	Econoburn EBW-150-O	Central Boiler E-Classic 1450	Portage & Main Optimizer 250
BTUs (Output)	150,000	214,000	180,000
Water Capacity	37 gallons	200 gallons	220 gallons
Efficiency	87%	90%	94%
EPA Phase 2 Certified	No	Yes	Yes
Cost	US$9,295	US$10,495	US$8,500

Optimizer 250 burns all types of seasoned wood and has a unique heat-exchanger design that passes superheated gases up to six times between two combustion chambers, allowing for maximum heat transfer into the water jacket, which is well insulated with R20 insulation. The resulting optimal high-temperature burn produces more heat from less wood.

- EPA Phase 2 certified
- Combustion chambers lined with heavy-duty, heat-treated refractory brick
- Digital controls are accurate to within one degree

Appendix
Tree Identification Guide

American Beech (*Fagus grandifolia*)

Tree: The shade-tolerant, broad-crowned American beech can reach a height of 100 feet (30 m) when mature.

Leaf: The ovate simple leaf is 2 to 5 inches (5–13 cm) long with up to 15 pairs of parallel veins. The upper surface is a glossy dark green; the underside is a paler green.

Bark: The American beech's smooth, thin bark is light bluish gray in color; it is susceptible to beech bark disease.

Black Cherry (*Prunus serotina*)

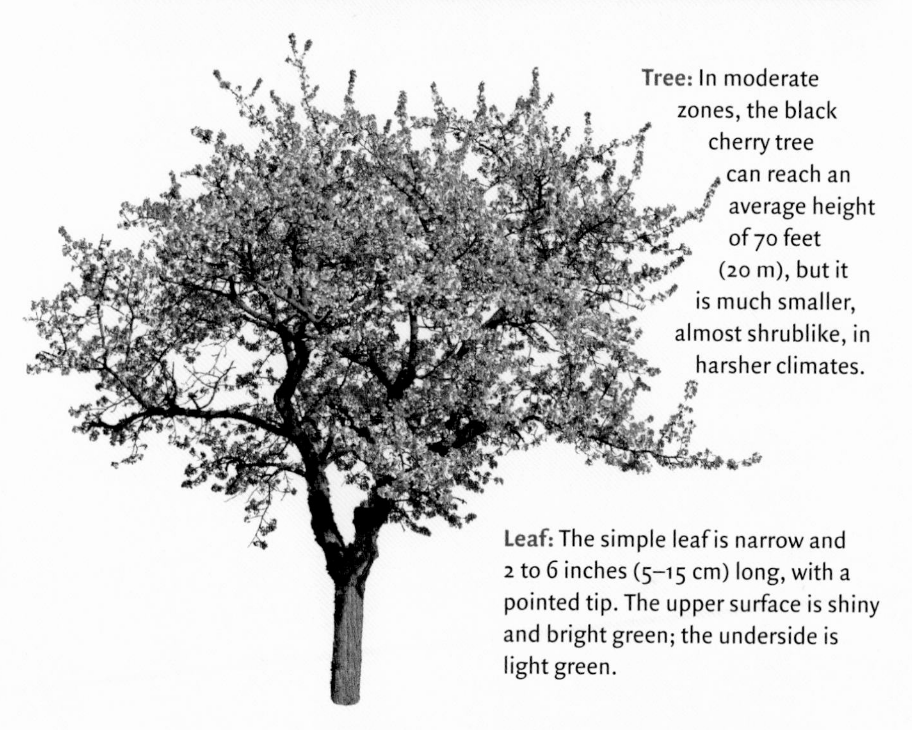

Tree: In moderate zones, the black cherry tree can reach an average height of 70 feet (20 m), but it is much smaller, almost shrublike, in harsher climates.

Leaf: The simple leaf is narrow and 2 to 6 inches (5–15 cm) long, with a pointed tip. The upper surface is shiny and bright green; the underside is light green.

Bark: The bark on a mature tree is dark gray, thick and extremely rough and scaly.

Black Locust (*Robinia pseudoacacia*)

Tree: The medium-sized black locust can reach a height of 70 feet (20 m). It has a narrow crown and an open, irregular form.

Leaf: With multiple small, paired leaflets under 2 inches (5 cm) long, the compound leaf can be up to 14 inches (36 cm) long. The upper surface is green; the underside is a paler green.

Bark: Mature bark is dark gray, thick and heavily ridged and furrowed.

Black Walnut (*Juglans nigra*)

Tree: A medium to large tree, the black walnut can reach a height of 100 feet (30 m). The trunk is straight with strong, large branches leading to a somewhat open crown.

Leaf: The large compound leaf is from 10 to 20 inches (25–50 cm) long and has multiple serrated leaflets, each about 2 to 4 inches (5–10 cm) long. The upper surface is yellow-green; the underside is slightly paler.

Bark: Mature bark is dark gray or dark brown in color and is ridged and furrowed in a rough diamond pattern.

Range

Eastern White Pine (*Pinus strobus*)

Tree: This large pine can reach a height of 100 feet (30 m). Its trunk is extremely straight, and the crown consists of upturned branches.

Leaf: Foliage consists of bundles of five soft needles that are 2 to 5 inches (5–13 cm) long and bluish green in color.

Bark: Brown to red-brown in color, with some gray, the young bark is thin, becoming thicker and darker as the tree matures.

Northern Red Oak (*Quercus rubra*)

Tree: The northern red oak is a medium to large tree that can reach a height of 90 feet (27 m). It has a rounded or irregular crown.

Leaf: The oblong leaf is 4 to 8 inches (10–20 cm) long, with multiple bristle-tipped lobes. The upper surface is dark green; the underside is paler.

Bark: As the smooth, dark gray bark matures, it develops dark, shallow furrows and gray ridges.

Red Maple (*Acer rubrum*)

Tree: The typically oval-shaped common red maple can reach a height of roughly 80 feet (25 m).

Leaf: The simple, opposite leaf is from 3 to 4 inches (7.5–10 cm) wide with three to five shallow lobes on each side. The upper surface is green; the underside is much paler.

Bark: The bark is a light gray color and smooth in texture on young trees, but as the tree ages, it becomes darker and rougher in texture.

Shagbark Hickory (*Carya ovata*)

Tree: The long-lived shagbark hickory tree reaches an average height of 70 to 80 feet (20–25 m). In open areas, its crown is oblong; in forested areas, the tree has a straight, slender crown.

Leaf: The compound leaf is from 12 to 24 inches (30–60 cm) long and typically has five leaflets. The upper surface is yellowish green; the underside is slightly hairy.

Bark: Its coarse, rough bark has long, loose, platelike strips that give the tree its shaggy appearance and name.

Range

Sugar Maple (*Acer saccharum*)

Tree: The sugar maple is a medium to large tree that can reach a height of up to 115 feet (35 m). It has a dense, rounded crown.

Leaf: The five-lobed leaf can grow to almost 8 inches (20 cm) wide. The upper surface is green; the underside is paler.

Bark: The young tree's bark is smooth and gray but splits into curling ridges as the tree ages.

White Ash (*Fraxinus americana*)

Tree: A medium-sized tree that can reach a height of up to 80 feet (25 m), the white ash typically has a tall, straight trunk with a narrow, oblong crown.

Leaf: The opposite, compound leaf is 8 to 12 inches (20–30 cm) long and typically has seven leaflets. The upper surface is green; the underside is slightly paler.

Bark: The bark is ash-gray to brown in color and features a ragged diamond pattern, with occasional scaly patches on older specimens.

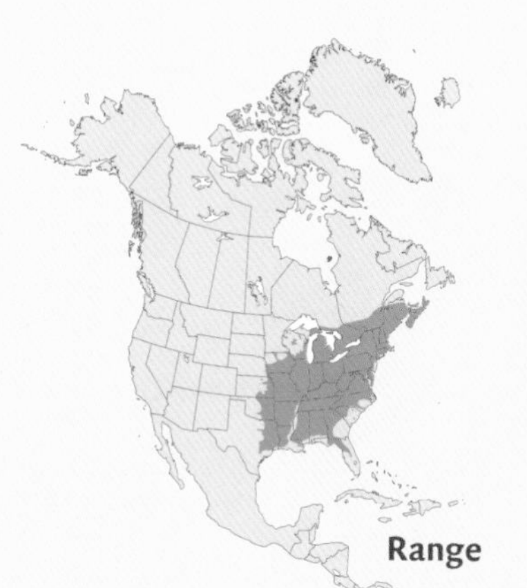

White Birch (*Betula papyrifera*)

Tree: This relatively short-lived tree grows to an average height of 50 feet (15 m), with one or several slender trunks and a small, open crown.

Leaf: The simple ovate leaf is serrated with a pointed tip and rounded base and is between 2 and 4 inches (5–10 cm) long. The upper surface is green; the underside is paler.

Bark: The bark is dark red on young stems but becomes creamy white and paperlike on mature trees, where it is easily peeled in horizontal strips.

White Oak (*Quercus alba*)

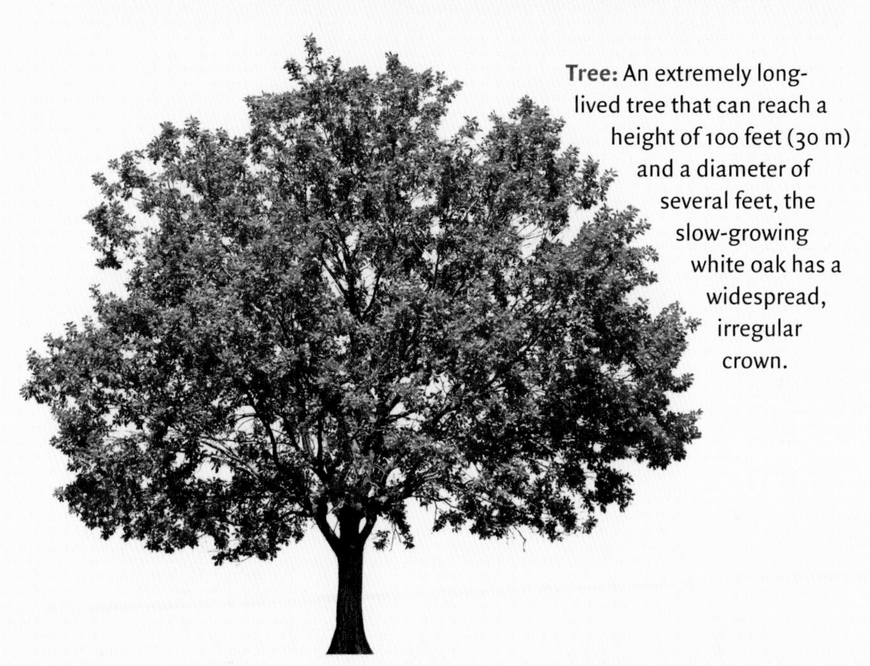

Tree: An extremely long-lived tree that can reach a height of 100 feet (30 m) and a diameter of several feet, the slow-growing white oak has a widespread, irregular crown.

Leaf: The leaf is simple, usually from 4 to 7 inches (10–18 cm) long with seven or more rounded lobes. The leaf tip is slightly rounded, and the base is wedge-shaped. The upper surface is green; the underside is white or gray.

Bark: The bark is whitish gray or gray in color with irregular patterns. Mature trees may feature smooth patches.

Photo Credits

Shutterstock

2 and cover: Mykhaylo Palinchak and Olga Nikonova; 8: Elena Larina; 12: Mykola Komarovskyy; 13 (top): Elena Elisseeva; 13 (bottom): Andreas Nilsson; 14: Semmick Photo; 16: Anna Moskvina; 25: Ewa Studio; 27: Irina Fischer; 28: OHishiapply; 33: by Paul; 37: Sascha Preussner; 41: adistock; 49: charnsitr; 51: Pshenina_m; 52: XXLPhoto; 57: ARENA Creative; 78: Nikonova Margarita; 89: Rob Bayer; 91: Eva Madrazo; 92 and spine: Jacek_Kadaj; 105: Jo Ann Snover; 106: Petar Ivanov Ishmiriev; 109: AdStock RF; 114: Maria Bobrova; 136: Buslik; 154: Johnny Adolphson; 156 (tree): Yuriy Kulik; 159 (tree): Jan Martin Will; 160 (tree): Denys Kurylow; 162 (tree): Le Do; 164 (tree): Melinda Fawver; 165 (tree): Maksym Bondarchuk; 166 (tree): vnlit; 167 (tree): LilKar; 168: mervas

iStock

157 (tree): Kerrick; 158 (tree): Antagain; 161 (tree): Kerrick; 163 (tree): DNY59

All appendix bark images courtesy of Bugwood.org

156: Joseph O'Brien, USDA Forest Service; 157: Bill Cook, Michigan State University; 158: William Fountain, University of Kentucky; 159: Bill Cook, Michigan State University; 160: Keith Kanoti, Maine Forest Service; 161: Bill Cook, Michigan State University; 162: John Ruter, University of Georgia; 163: Rob Routledge, Sault College; 164: Rob Routledge, Sault College; 165: Keith Kanoti, Maine Forest Service; 166: Keith Kanoti, Maine Forest Service; 167: Richard Webb, Self-employed horticulurist

Miscellaneous

129: Elmira Stove Works

Resources

CANADA

Association des professionnels du chauffage (A.P.C.)
Web: www.poelesfoyers.ca
Tel: (450) 748-6937

Canada Mortgage and Housing Corporation (CMHC)
Web: www.cmhc-schl.gc.ca

Canadian Standards Association (CSA)
Web: www.csagroup.org

Hearth, Patio & Barbecue Association of Canada
Web: hpbacanada.org
Tel: (705) 788-2221
E-mail: hpbac@bellnet.ca

Insurance Bureau of Canada
Web: www.ibc.ca/en/Contact_Us.asp

Wood Energy Technology Transfer Inc. (WETT)
Web: www.wettinc.ca
Tel: (416) 968-7718 / (888) 358-9388
E-mail: info@wettinc.ca

UNITED STATES

Chimney Safety Institute of America (CSIA)
Web: www.csia.org
Tel: (317) 837-5362

Masonry Heater Association of North America
Web: www.mha-net.org
Tel: (520) 883-0191
E-mail: execdir@mha-net.org

National Fireplace Institute (NFI)
Web: http://nficertified.org
Tel: (703) 524-8030
E-mail: info@nficertified.org

Pellet Fuels Institute
Web: http://pelletheat.org
Tel: (703) 522-6778
E-mail: pfimail@pelletheat.org

U.S. Environmental Protection Agency (EPA)
Web: www.epa.gov

Wood Heat Organization Inc.
Web: woodheat.org
E-mail: question@woodheat.org

Index

Z